TRONNIES

THE SOURCE OF THE COULOMB FORCE
AND
THE BUILDING BLOCKS
OF UNIVERSES

THE ROSS MODEL
A THEORY OF EVERYTHING
VERSION EIGHT
BY JOHN R. ROSS

Illustrations by Marshall Ross

Published by John R. Ross
P.O. Box #2138
Del Mar, CA 9014

Second Printing

TABLE OF CONTENTS

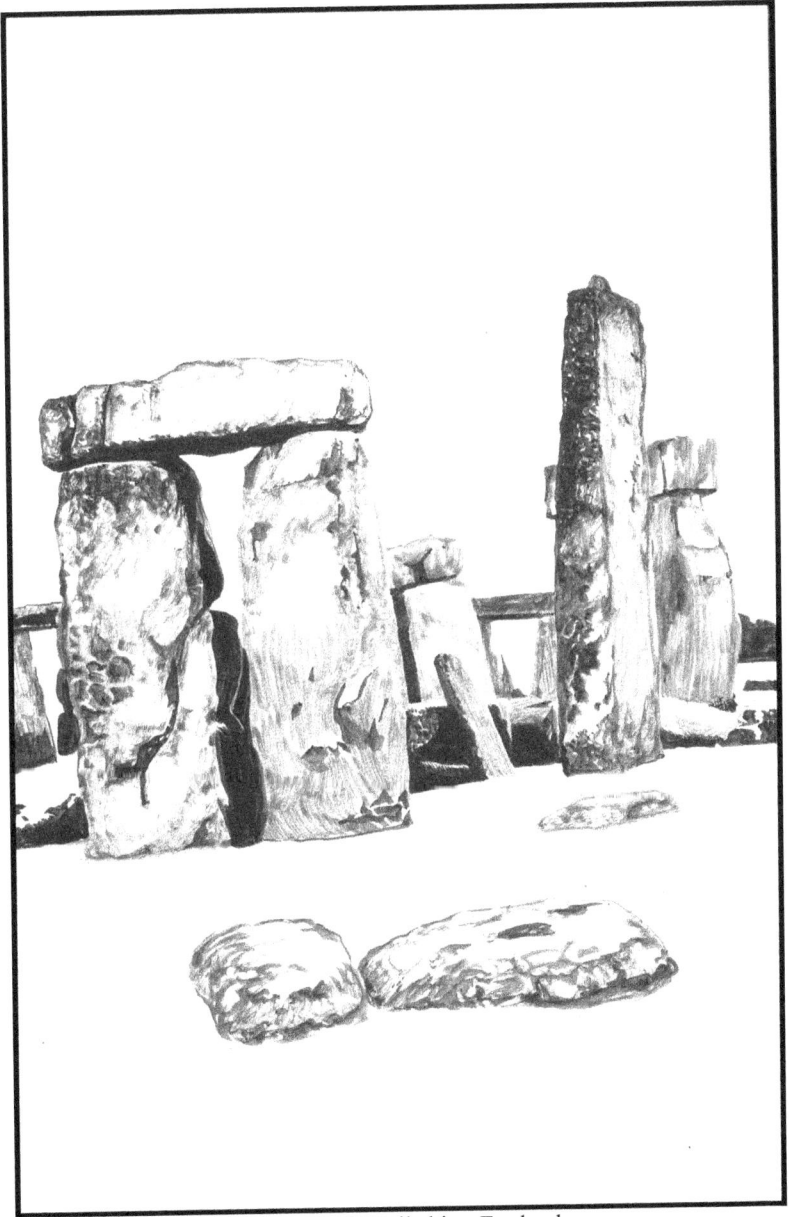

Stonehenge, Wiltshire, England

Mr. Ross has asked me to provide a short introduction to and summary of each of his chapters with emphasis on the main issues discussed in the chapters.

Chapter I is a short summary of the generally accepted scientific beliefs and theories as to how our universe was created and evolved. This is what you should have learned in your science classes. Since most of you will have forgotten much of the details, you may appreciate this short summary. In any case when you have finished Chapter I, you should have a feel for the current scientific thinking about our Universe and its 100 to 400 billion galaxies and then it will be time to try to digest the Ross Model of our Universe.

CHAPTER I

ACCEPTED SCIENTIFIC BELIEFS ABOUT OUR UNIVERSE

Search for the Truth

Since the beginning of human civilization mankind has searched for explanations of the origin of our Universe, how it was put together and how it works. Early efforts involved supernatural beings and religions evolved from these efforts. More recent explanations have involved complicated mathematical models based on experimental evidence, some involving multiple extra dimensions. Many millions of dollars are being spent in the United States alone and similar efforts are underway in other countries in search of the ultimate tiny building blocks of our Universe and a process for explaining all of nature.

Theories of Everything

A single theory that explains everything in our Universe and everything about our Universe is called a "theory of everything". The famous physicist, Stephen Hawking wrote a book entitled "*The Theory of Everything*" that is an excellent summary of current scientific understandings about our Universe, but he ended his book with an admission that science had become too complicated for philosophers and they had ceased asking questions such as "Did the universe have a beginning?" and he concluded his text as follows:

However, if we discover a complete theory, it should in time be understandable in broad principal by everyone, not just a few scientists. Then we shall be able to take part in the discussion of why the universe exists. If we find the answer to that, it would be the ultimate triumph of human reason. For then we would know the mind of God.

3

Popular Scientific Models

I believe my model of our Universe is the complete theory that Dr. Hawking was contemplating, but before I provide the details of my model of our Universe, I will provide a short summary of currently popular scientific models. These popular scientific models propose a complicated set of so-called "elementary particles" that are supposed to be building blocks of matter. These particles include electrons, positrons, six types of quarks (three of which are supposed to make a proton and three of which are supposed to make a neutron) and neutrinos. Neutrinos are supposed to be produced in the sun, have the same spin as the electron, travel at, or very close to, the speed of light and most of them that illuminate the earth, according to accepted theories, pass right through it. Scientists do not have a good explanation of the internal structure of electrons or protons. Scientists are not able to isolate any of the six quarks or convincingly prove they exist. Scientists have detected flashes of energy in underground tanks that they attribute to neutrinos but there is no good proof that neutrinos exist.

These popular models also include a complicated set of forces. These include electromagnetic forces (that combines Coulomb electric forces with magnetic forces), the "strong" force which is supposed to hold atomic nuclei together, a "weak" force related to a type of radioactive decay and the force of gravity. Our scientists have learned to deal very effectively with these forces but are unable to provide a deep understanding of the sources of these forces. Those models include a variety of electromagnetic radiation, including gamma rays, x-rays, ultraviolet light, visible light, infrared light, millimeter waves, microwaves and radio waves.

Big Numbers and Little Numbers

Right now I need to digress a bit to make sure you readers remember what you learned about how to deal with extremely large numbers and extremely small numbers utilizing exponents. If we are going to talk about galaxies and the internal structure of electrons and protons, we will be dealing with a very wide range of numbers and values. Scientist and engineers have developed a trick for doing this using the number 10 and an exponent. I have provided some simple examples in Table I.

For example the speed of light is about 299 million meters per second. We can write it a 2.99×10^8 m/s. In the hydrogen-1 atom an electron orbits a proton at a distance of about 0.0000000000529 meters. We can write that orbit radius as $r = 0.529 \times 10^{-10}$ m.

Table I
Very Big and Very Little Numbers

Big and Bigger
$10^0 = 1 =$ one
$10^1 = 10 =$ ten
$10^2 = 10 \times 10 = 100 =$ one hundred
$10^3 = 10 \times 10 \times 10 = 1,000 =$ one thousand
$10^4 = 10 \times 10 \times 10 \times 10 = 10,000 =$ ten thousand
$10^6 =$ one million
$10^9 =$ one billion
$10^{12} =$ one trillion
$10^{15} =$ one thousand trillion
$10^{18} =$ one million trillion

Little and Littler
$10^0 = 1 =$ one
$10^{-1} = 1/10 = 0.1 =$ one tenth
$10^{-2} = 1/(10 \times 10) = 1/100 = 0.01 =$ one hundredth
$10^{-3} = 1/(10 \times 10 \times 10) = 1/1,000 = 0.001 =$ one thousandth
$10^{-4} = 1/(10 \times 10 \times 10 \times 10) = 1/10,000 = 0.0001 =$ one ten thousandth
$10^{-6} =$ one millionth
$10^{-9} =$ one billionth
$10^{-12} =$ one trillionth
$10^{-15} =$ one thousand trillionth
$10^{-18} =$ one million trillionth

Our Universe Is Very Large

Our earth and its sun are part of the Milky Way Galaxy. The Milky Way is an average size galaxy that has in it about 100 billion stars. (Our sun is one of those stars.) There are in our Universe more than 100 billion galaxies, perhaps about 400 billion galaxies. This means there are more than (100 $\times 10^9$ galaxies) $\times$ (100 $\times 10^9$ stars/galaxy) = 10^{22} stars in our Universe. That is 10 trillion billion stars, 40 trillion billion if you use the 400 billion galaxies estimate. Many if not almost all of these stars are believed to have planets orbiting them and many if not most of these planets are likely to have moons. The distance to the edge of the observable universe is about 10^{26} meters. That is 100 trillion-trillion meters. We know that our Universe is a very big place with lots of stuff in it, but scientists do not have a good explanation of how it was created or what ultimately is going to happen to it.

The Birth of Our Universe – The Big Bang

We also know that our Universe is currently expanding with the distances between far away galaxies (or clusters of galaxies) expanding faster depending on how far away the galaxies (or clusters) are from each other. We estimate the age of our Universe by dividing the distance to each far away galaxy by the speed at which the galaxy is moving away from our galaxy, the Milky Way. We get roughly the same answer for each galaxy for which we have a good estimate of distance. Based on these measurements and calculations scientists believe that our Universe originated about 13 to 15 billion years ago in a Big Bang explosion. Some scientists have predicted its birth date more precisely at "approximately 13.72 billion years ago. About 300,000 years after the Big Bang a very large number of small atoms, mostly hydrogen atoms and a much smaller number of helium atoms had formed. Over time these atoms collected into gas clouds that later became stars. In the extreme heat in the core of stars hydrogen atoms combined to produce helium releasing energy in the process, and hydrogen and helium combined to produce larger atoms and these larger atoms combine with other atoms to make even larger atoms. Stars collected into galaxies. Some stars exploded sending atomic debris spreading out into interstellar space. Planets, including our earth, formed from collections of the atomic debris of exploded stars. Some planets are mostly hot gases but scientists believe there are many planets with conditions similar to conditions on earth that are capable of supporting the development of life.

Little Things Are Still Not Described

Scientists know a lot about our Universe but scientists after centuries of scientific progress still do not have a good understanding of what goes on at sub-atomic levels or a good description of sub-atomic things such as electrons, protons and photons. We utilize electricity and magnetism every day but we do not have a good understanding of what electricity and magnetism really are. Scientists also do not have a good understanding of the force (the so-called "strong" force) holding atomic nuclei together or the "weak" force involved in a type of nuclear decay. They know a lot about the Coulomb force. This is a force between electric charges that is repulsive for like charges and attractive for unlike charges. The force gets greater as the charges get closer together according to Coulomb's Law. Coulomb's law teaches us that the repulsive force between like charges touching each other should be infinite. Scientists cannot explain why the electron with a single charge and a size close to zero does not blow itself apart with its own Coulomb force.

Big Things are Still Not Understood

Scientists know that close-by galaxies are attracting each other and far-away galaxies are receding from each other, but they do not know why. There is no agreement as to what preceded the Big Bang. Scientists do not know whether our Universe will continue to expand forever or if it will at some time begin to contract. There is no answer to the question of whether our Universe is finite or infinite, or if it is finite what forms the boundary of our Universe.

Atoms

There are 92 types of naturally occurring atoms, each with a nucleus and a unique number of associated electrons. Atoms each have a single relatively heavy positively charged nucleus and the nucleus has associated with it one or more electrons, each of which has a negative charge of -e. The number of associated electrons in a charge neutral atom represents the atomic number of the atom. Some of the more familiar atoms are listed in **Table II**.

The net charge of the nucleus of each neutral atom is equal and opposite the total charge of the number of its electrons associated with the nucleus. So for example the net charge on the helium nucleus is + 2e and the net charge of its two orbiting electrons is – 2e. The net charge of the helium atom is 0 (since +2e added to -2e = 0). The copper nucleus has a net charge of plus 29 and it has 29 electrons associated with its nucleus, 28 of which orbit the nucleus. In a piece of copper metal one of those electrons is a conduction electron and is free to move freely in the matrix of copper atoms making up the piece of the copper metal.

Table II
Typical Atoms

Number of Associated Electrons	Atom	Symbol
1	Hydrogen	H
2	Helium	He
6	Carbon	C
7	Nitrogen	N
8	Oxygen	O
10	Neon	Ne
11	Sodium	Na
12	Magnesium	Mg
13	Aluminium	Al
14	Silicon	Si
16	Sulfur	S

18	Argon	Ar
20	Calcium	Ca
26	Iron	Fe
29	Copper	Cu
47	Silver	Ag
79	Gold	Au
82	Lead	Pb
92	Uranium	U

Electrons

There are two types of electrons:

(1) the type most people are familiar with that has a negative charge of –e (its official name is "negatron" but it is usually referred to as an "electron") and

(2) the type most people are not familiar with that has a positive charge of +e. It is called a "positron". The positron is the anti-particle of the negative electron. This anti-particle is exactly like the negative electron except for its positive charge of +e.

Pair Production and Electron-Positron Annihilation

Pairs of electrons (one negatron and one positron) can be produced when high-energy photons (called gamma rays or gamma ray photons) interact with matter. When an electron and a positron combine they both vanish and are replaced by high-energy photons. These processes are respectively called "pair production" and "electron-positron annihilation". Existing theories do not provide a good explanation of pair production or electron-positron annihilation.

Einstein ushered in modern physics with his landmark papers on relativity, energy and matter. The Ross Model is inconsistent with many of his explanations.

Photons

Visible light is a type of radiant energy that allows us to see. The types of radiant energy

8

that are a part of an electromagnetic spectrum which (with decreasing energy) includes gamma rays, x-rays, ultraviolet light, visible light, infrared light, microwaves and radio waves. For more than 100 years scientists have known that this radiation energy comes in separate and distinct "quantities" of energy. Albert Einstein is often credited with identifying the quantum nature of the photon. These separate and distinct quantities of energy are called photons. But scientists still, after 100-plus years, do not know what a photon is or what it looks like. All of this radiation travels at the speed of light. The speed of light is about 3×10^8 meters per second (about three hundred million meters per second).

Photon energy is usually expressed in joules (J), newton-meters (Nm) or electron volts (eV). The energy of a photon, E_{photon}, can be calculated from its wavelength λ or frequency f using one of the following formulas:

$$E_{photon} = hc/\lambda = hf \qquad (1)$$

where h is Planck's constant = about 6.626×10^{-34} Nms = about 6.626×10^{-34} Js = about 4.135×10^{-15} eVs.

N is the symbol for newton, a unit of force, m is meters, eV is electron-volt, a tiny unit of energy and s is seconds. A newton-meter, Nm is the same as a joule, J, a very much larger unit of energy. The symbol c is the speed of light, and λ is the wavelength (usually expressed as a very small fraction of a meter) of the photon, so for example: the wavelength of a green light photon is about 5.4×10^{-7} m, so the energy of this green light photon is:

$$E = (6.626 \times 10^{-34} \text{ Js})(3 \times 10^8 \text{ m/s})/(5.4 \times 10^{-7} \text{ m}) = 3.68 \times 10^{-19} \text{ J}$$
$$= 2.3 \text{ eV}$$

The energy of a single photon of visible light is an extremely small amount of energy, but you should realize that when you are looking at a tree leaf, even at a distance or several meters, thousands or millions or billions of these green light photons are being focused by the cornea and lens of your eyes onto your retina each second so you can see the tree and visualize its green leaves. The smaller the wavelength of the photon the larger is its energy. Radio wave photons have relatively long wavelengths and gamma ray photons have relatively short wavelengths. Visible light photons are somewhere in the middle.

Mass, Energy, Conversion Units and Universal Constants
Existing reference books contain precisely measured values of the masses of sub-atomic

particles and their equivalent energy and provide precise values of important conversion units and universal constants such as the electron charge and the speed of light.

Some of these values needed to understand portions of the Ross Model are listed in **Tables III, IV and V.** To follow some of the math in the following chapters, readers may want to refer from time to time to the values in these tables. For now many readers will find the tables complicated. For those that do, I suggest you simply skim the tables to check your memory as to which of the information you already know.

Table III
Masses of Some Small Atoms and Particles

Particle or Atom	Symbol	Mass (kg)	Energy (MeV)
Electron at rest	e-	$9.109\ 3897 \times 10^{-31}$	0.510 712 57
Positron at rest	e+	$9.109\ 3897 \times 10^{-31}$	0.510 712 57
Proton	p	$1.672\ 6231 \times 10^{-27}$	938.272 338
Neutron	n	$1.674\ 9286 \times 10^{-27}$	939.565 628
Deuteron	d	$3.343\ 5860 \times 10^{-27}$	1875.613 39
Tritium isotope	^3H	$5.008\ 2711 \times 10^{-27}$	2807.857 70
Hydrogen one atom	^1H	$1.673\ 5340 \times 10^{-27}$	938.256 992
Helium 4 atom	^4He	$6.646\ 4835 \times 10^{-27}$	3726.311922

Table IV
Some Important Conversion Units

One electron volt	eV =	$1.602\ 177\ 33 \times 10^{-19}$ J	joules
	eV =	$1.783\ 662\ 70 \times 10^{-36}$ kg	kilograms
	eV =	4.45×10^{-26} kWhr	kilowatt-hours
	eV =	$1.602\ 177\ 33 \times 10^{-19}$ Ws	watt-second
eV/molecule		eV/molecule = 96.49 kJ/mole	kilo-joules per mole
kJ/mol		kJ/mol = 1.0365×10^{-2} eV/molecule	eV/molecule
Kcal/mol		Kcal/mol = 4.336×10^{-2} eV/molecule	eV/molecule
One atomic mass unit	amu =	$1.660\ 5402 \times 10^{-27}$ kg	kilograms
	amu =	932.0 MeV	million electronvolts
One kilogram	kg =	$8.987551787 \times 10^{16}$ J	joules
Joule (energy)	J =	kgm^2/s^2	kilogram meter squared per second squared

Joule	$J = C^2/F$	coulomb squared per Farad

Newton (force)	$N = kgm/s^2$	kilogram meter per second squared
Watt (power)	$W = J/s$	joule per second

Standard Temperature $25°C = 298K = 77°F$

Year	$Yr = 3.1536 \times 10^7 \, s$	seconds

Light year	$ly = 9,460,730,472,580,800 \, m$ $= 9.46 \times 10^{15} \, m$	meters

Temperature (degrees)	$F = (K - 273)1.8 + 32,$	where F is Fahrenheit, K is Kelvin
	$F = 1.8C + 32$	where F is Fahrenheit, C is Centigrade
	$°C = K - 273°$	

Volume $V = (4/3)\pi r^3$

Table V
Universal Constants

Speed of light in vacuum	$c = 2.99\,792\,458 \times 10^8 \, m/s$	meters per second
Plank constant	$h = 6.626\,0755 \times 10^{-34} \, Js$ $h = 4.135\,6692 \times 10^{-15} \, eVs$ $\hbar = 1.054570203 \times 10^{-34} \, Js$	joule-second electron-volt second joule-second
hc	$hc = 19.86447461 \times 10^{-26} \, Jm$ $hc = 12.38942435 \times 10^{-7} \, eVm$	joule-meter electron-volt-meter
Avogadro constant	$N_A = 6.022\,1367 \times 10^{23}/mole$	per mole
Vacuum permittivity	$\varepsilon_o = 8.854188 \times 10^{-12} \, C^2/Nm^2$	coulomb squared per newton meter squared
Coulomb constant	$k = 8.987551787 \times 10^9 \, Nm^2/C^2$	newton meter^2/coulomb^2
Boltzmann constant (particle energy per K)	$k = 8.62 \times 10^{-5} eV/K$ $k = 1.38 \times 10^{-23} J/K$ $k = 1/4\pi\varepsilon_o$	electron-volts/degree kelvin joule/degree kelvin coulomb squared per newton meter squared

Pi	π = about 3.1416...	
Electron Charge/ Elementary charge	e = 1.602 177 33 X 10^{-19} C	coulombs
Coulomb	C = 6.2415 X 10^{18} e	electrons
Ampere	Amp = 1 C/s	coulomb per second
	Amp = 6.24 X 10^{18} e/s	electrons/second
Farad	F = C^2/J = C^2/Nm	coulomb squared per Joule
Wien's Law	λ = 2.898 X 10^{-3} mK/T	
	E = 6.86 X 10^{-23} J-T/K	
	= (4.28 X 10^{-4} eV-T/K)	

where λ is the peak wavelength of radiation emitted from a black body at temperature T in degrees Kelvin (K). E is energy of radiated photons. J is joules and eV is electron-volts which are energy units.

The electrical force (also called the "Coulomb Force") F, between stationary charged particles is:

$$F = kQ_1Q_2/r^2, \qquad (2)$$

where k is the Coulomb constant, k = 8.99 X 10^9 Nm^2, Q_1 and Q_2 are the charges (in Coulombs) of the particles and r is the distance between the particles.

Orbital velocity of a satellite: $V_G = \sqrt{\dfrac{GM}{r}}$

where M is the satellite mass, r is the orbit radius and G is the gravitational constant = 6.67428 $X10^{-11}$ Nm^2/kg^2.

Avogadro's constant from **Table IV** represents the number of atoms of a particular material in a number of grams equal to the atomic mass number of the material. Pi (π) from **Table IV** is the ratio of the circumference of a circle to the circle's diameter. Plank's constant from **Table IV** gives us the energy of a photon using **Equation (1)** if we know its wavelength. Boltzmann constant times the temperature of a black body in degrees kelvin gives the peak energy of radiation from a radiating body.

Table VI
Estimated Mass and Size of Big Things

	Mass	Diamter
Earth	5.98×10^{24} kg	$\sim 1.27 \times 10^{7}$ m
Sun	1.59×10^{30} kg	$\sim 1.39 \times 10^{9}$ m
Milky Way Galaxy	2×10^{42} kg	$\sim 1.04 \times 10^{21}$ m
Our Universe	1.46×10^{53} kg	at least 9.2×10^{26} m

Good morning, Ladies and Gentlemen and Boys and Girls. This Chapter II is a quick summary of the Ross Model of our Universe. We have included it at the beginning just to give you a good feeling about what you might expect in the rest of the book.

In a nutshell, tronnies are point particles with a charge of plus e or minus e. Tronnies have no mass and no volume. They are points. Everything in our Universe is made from tronnies or things made from tronnies.

All tronnies travel in circles in twosomes or threesomes at speeds of about 1.57 times the speed of light (exactly $\pi/2$ times the speed of light).
The twosome is a particle called an "entron". It is comprised of a plus tronnie and a minus tronnie, has no net charge, but has a mass-energy (determined by the diameter of its circle. As far as we know Mr. Ross was the first person in our Universe to identify tronnies and entrons.

The threesomes are either "electrons", each one comprised of two minus tronnies and one plus tronnie, or "positrons" each one comprised of two plus tronnies and one minus tronnie. A positron is the anti-particle of the electron. So entrons, electrons and positrons are made from tronnies. Everything else in our Universe is made from entrons, electrons and positrons. By "everything", we mean everything.

CHAPTER II

THE ROSS MODEL
(A Quick Summary)

I have developed a "theory of everything"; I call it the "Ross Model". This **Chapter II** is a very brief summary of my entire theory. The details are spread out in the following chapters which I will reference in this chapter. Unlike Albert Einstein's theories of relativity and much of the current theories describing things such as subatomic matter, light, gravity, electricity, magnetism, black holes and the beginning and end of our Universe; my theory is simple and easy to understand, at least in broad principle, by everyone. **I must warn you that my theory is a theory, and it is certainly not generally accepted by the current scientific community.** The basic concept of my theory is that everything in our Universe (including photons, electrons, atoms, molecules, gravity, stars, planets, moons and us) is made from a single type of point particle with an electric charge of plus e and its anti-particle with an electric charge of minus e, each of which has no mass and no volume and they are always traveling faster than the speed of light! To the best of my knowledge, I was the first person in our Universe to recognize the existence of these particles and have named them, **"tronnies"**.

My theory is inconsistent with a generally accepted scientific model called the "Standard Model" that proposes a wide variety of basic building blocks of our Universe. My theory is also inconsistent with Einstein's theories of relativity, which are generally accepted as correct but only a handful of scientists really understand them. My theory solves the same problems that Einstein solved, but my theory can be understood by everyone. My intention with this book is to explain my theory as simply as possible. I want everyone to have the opportunity to know how our Universe was created and how it functions at the most basic levels. I will try to explain my theory so that it can be understood by anyone who has at least the knowledge of science of a typical smart child who has graduated from **junior high school**. Professor Trebla Nietsnie ("Albert Einstein" spelled backwards) will introduce each chapter and simplify the principal features of my theory. I also want people with doctoral degrees in physics, chemistry and astrometry to examine and understand the specific details of my theory. I expect major criticism from the scientific community and to the extent there are errors in my theory, I welcome criticism and especially suggestions for corrections. However, accusations of errors should be based on facts and experiments not on existing theories, many of which are clearly inconsistent with the Ross Model.

Generally Accepted Scientific Beliefs

Most of you readers will have forgotten most of what you learned in school about science. Therefore, as you have just discovered, I have provided, as a reminder in **Chapter I** a brief description of accepted scientific beliefs about our Universe including a very short summary of current popular scientific theories describing what scientists believe are the elementary particles from which our Universe is constructed. The birth of our Universe and its size is also discussed briefly in **Chapter I** along with current descriptions of photons, atoms, molecules, stars, planets and moons. I have also included in **Chapter I** tables giving masses and energies of particles such as electrons and protons and a set of Universal constants and conversion units that are needed to support some of my calculations later on in the book. Also I have included two tables which will allow people who have forgotten how to handle exponents to more easily deal with extremely small and extremely large numbers.

Tronnies

The star characters in this book are the tronnies. I discovered tronnies a few years ago and gave them their name. Tronnies are the basic building blocks of our Universe.

Everything in our Universe is made from tronnies or things that are made from tronnies. Tronnies have no mass and no volume; they are point particles.

Each tronnie is a point focus of a type of force called the "Coulomb force" which travels as plus or minus waves at the speed of light. This point focus of Coulomb plus or minus forces gives each tronnie a "charge" which is either plus or minus. The Coulomb forces converge to the point from all directions at the speed of light and they expand out from that point in all directions at the speed of light. There are an equal number of plus and minus tronnies in our Universe. The minus charge is equivalent to the net charge of an electron and the plus charge is equivalent to the net charge of a proton or a positron.

Tronnies, because of their charges, are attracted and repelled by Coulomb forces. Minus tronnies attract plus tronnies and repel other minus tronnies. Plus tronnies attract minus tronnies and repel plus tronnies. This repulsion and attraction is at the speed of light! *Every tronnie, being exactly like itself, repels itself with its own Coulomb force waves that expand at the speed of light!* Tronnies pair up with other tronnies in twosomes and threesomes with each tronnie traveling in a perfect circle at a speed of $\pi c/2$ (approximately 1.57 times the speed of light). Two tronnies (one plus and one minus) traveling on opposite sides of a perfect circle is a particle called an "**entron**". (Like the tronnie, I discovered and named the entron.) Two minus tronnies and one plus tronnie combine to form an "**electron**" and two plus tronnies and one minus tronnie combine to form a "**positron**". (The electron and the positron were discovered long ago, but I was the first to describe their structure.) Tronnies are discussed and described in detail in **Chapter III**. Tronnies ride their own Coulomb force waves similarly to a surfer riding an ocean wave, which means that the tronnie can travel faster than the speed of light but not slower. A drawing of a tronnie repelling itself along the arc of a circle faster than the speed of light is shown in **FIG. 1** in **Chapter III**.

Entrons

The smallest quantity of anything is referred to by scientists as a "**quantum**". The energy quantum of our Universe is the entron. Entrons are discussed in detail in **Chapter IV**. Each entron is comprised of one plus tronnie and one minus tronnie, each traveling on the opposite sides of a perfect circle at $\pi c/2$ (approximately 1.57 times the speed of light). According to the Ross Model, entrons represent all of the energy of our Universe and (except for the masses of electrons and positrons) all of the mass of our Universe. The energy of an entron and its mass are inversely proportional to the diameter of its circle. The size of the largest (lowest energy) entron is about 100 million billion times larger than the smallest entron. Entrons provide electrons their voltage. Entrons represent temperature and heat of matter and are released from matter in the form of photons. So there are infrared entrons, visible light entrons, ultraviolet entrons, X-ray entrons and gamma ray entrons. The smallest entron with the greatest energy and mass is the "**neutrino entron**". Each proton contains one neutrino entron which represents almost all of the mass of the proton. Neutrino entrons are released in Black Holes with the destruction of protons to provide the **gravity** of galaxies.

Photons

Each photon is a quantum of electromagnetic radiation. Electromagnetic radiation includes a wide spectrum of radiation such as radio wave radiation, microwave radiation, infrared radiation, visible light radiation, ultraviolet radiation, x-ray radiation and gamma ray radiation. Photons travel at the speed of light, about three hundred million meters per second (3×10^8 m/s).

Each photon is comprised of one entron which, as explained above, is two tronnies traveling in a circle at about 1.57 times the speed of light. In a photon the entron travels in a much larger circle at a speed of **twice** the speed of light (about 6 X 10^8 m/s) and forward at the speed of light. The diameter of the photon's circle is proportional to the diameter of the entron's circle and the mass of the photon is equal to the mass of its entron. (Most scientists believe that photons do not have mass, but photons do have mass as you will discover later on.) The diameter d of the photon circle is equal to $2/\pi$ (about 0.6366) times the wavelength of the photon which is the distance traveled forward by the entron in one of its cycles around the photon's circle. The diameter d' of the photon's entron is equal the photon's diameter d divided by 911 (thus d' = d/911). Photons are discussed in **Chapter V** and drawings showing the internal structure of a photon are shown in **FIGS. 3** and **4**. Please make note of the photon's circle and the entron's circles within the photon. The Ross Model describes a previously unknown but extremely important photon that I call the "**neutrino photon**" which is produced in "**Black Holes**" with the destruction of protons and provides the gravity of our Universe. Neutrino entrons and neutrino photons are discussed in **Chapter VI**. Gravity and Black Holes are discussed in **Chapter XXI**.

Electrons and Positrons

Most Americans are aware of electrons. Most Americans have never heard of positrons or if they have, they have forgotten what they are. You are about to learn a lot about positrons. Positrons are the anti-particle of the electron. I have provided a short description of what scientists currently know about positrons in **Chapter I**. There is much that scientists do not know about positrons. For example, the number of positrons in our Universe is exactly equal to the number of electrons in our Universe. There are two types of electrons and two types of positrons. These are "naked electrons" and "energetic electrons" and "naked positrons" and "energetic positrons".

Naked electrons and naked positrons are zero energy electrons and positrons. Each is comprised of only three tronnies. All naked electrons are exactly alike and all naked positrons are exactly alike. Naked positrons are exactly opposite naked electrons and they are referred to as the anti-particles of each other. The three tronnies of the naked electron and the naked positron circle in circles at speeds faster than the speed of light with frequencies of about 160 trillion trillion (1.6 X 10^{26}) cycles per second! This is called spinning! Two minus tronnies and one plus tronnie make a naked electron with a net charge of minus e and two plus tronnies and one minus tronnie make a naked positron with a net charge of plus e. Electrons and positrons are described in **Chapter VII** and a drawing of a naked (zero energy) electron is shown in **FIG. 5**. A drawing of a naked positron would look exactly like the drawing of the naked electron except the plus tronnie would be a minus tronnie and the two minus tronnies would be two plus tronnies. As you can see from the drawing their shapes are complicated but they each could fit in a cubic box with sides of 2×10^{-18} m. Except for tronnies that have no size at all and neutrino entrons, naked electrons and naked positrons are the smallest things in our Universe. They are about 100 million times smaller than a typical atom. (Niels Bohr figured out the speed of the electron in the hydrogen atom more than 100 years ago but he did not realize that the electrons were self-propelled.) Naked electrons and naked positrons are self-propelled at a speed of about 2.19 million meters per second (2.19×10^{6} m/s). Electrons and positrons capture entrons to become energetic electrons and positrons. Since the largest (lowest energy) entron is about 100 million billion times larger than the smallest entron and 100 million billion times less energetic and less massive, there can be a great deal of variations in the energy of energetic electrons and a great deal of variations in the energy of energetic positrons. An energetic electron is shown in **FIG. 6**.

Entrons, electrons and positrons are made from tronnies. Protons are made from an electron and two positrons (giving the proton its charge) and a neutrino entron (giving the proton almost all of its mass).

Protons

A naked proton is comprised of a naked electron, which has captured a neutrino entron (to become a **very energetic electron** with an increase in its mass of more than 1800 times) and two naked positrons. The neutrino entron has a mass of about 1.6564×10^{-27} kg and a frequency equal to the frequency of the naked electron and the two naked positrons. The neutrino photon has a diameter of about 0.85×10^{-15} m and its neutrino entron has a diameter of about 0.934×10^{-18} m (0.85×10^{-15} m divided by 911) which gives the neutrino entron a diameter which is about one-half the size of the naked electron. The neutrino entron forces the capturing electron into a circle the size of the neutrino photon (i.e. about 0.85×10^{-15} m). This defines the proton size, and the mass of the neutrino entron (i.e. about 1.6564×10^{-27} kg) provides the proton with almost all of its mass. Naked protons are self-propelled at speeds of more than 10 percent of the speed of light, but they collect gamma ray entrons to slow down enough to be able to collect an electron to become a hydrogen atom. Some of these gamma ray entrons are released when hydrogen atoms are fused to make helium in hydrogen bomb explosions and to make helium in fusion reactions in stars. Protons are discussed in more detail in **Chapter VIII** and a drawing of a naked proton is shown in **FIG. 7**.

Atoms

Atoms are fabricated in stars. Nature's technique for the construction of atoms is discussed in **Chapters XII** and **XIII**. Alpha particles are the nuclei of helium-4 atoms. An alpha particle is the combination of four protons (which provide almost all of the helium 4 mass) and two electrons plus some gamma ray entrons. Some of the gamma ray entrons of the hydrogen atoms are released in the course of the fusion of the four hydrogen atoms. We see this release of entrons as fusion energy. A graphical model of an alpha particle is shown in **FIG. 11** in **Chapter XII**. You will note that the alpha particle has a positive charge on its inside (plus 4) due to its four circling protons and a negative charge mostly on its outside (minus 2) due to its two electrons circling through and around the circular path of the four protons. Also the heavy portion of the alpha particle (the four circling protons) is on the inside and the light portion of the alpha particle is mostly on the outside (the two electrons). This gives each alpha particle a net charge of plus 2, which you might think would cause alpha particles to repel other alpha particles by reason of the repulsive Coulomb forces between the two net positively charged particles. They do repel each other at distances that are large compared to the size of the alpha particle. However, when alpha particles are close together, the heavy positive portions of the alpha particles are attracted to the light negative portion of its neighbor alpha particles and in particular configurations these attractive forces can slightly trump the net repulsive forces between the alpha particles. Therefore, three or more alpha particles, each with a mass number of 4, can combine to form the nuclei of heaver atoms. For example the nucleus of: carbon-12 contains three alpha particles, oxygen-16 contains four alpha particles and neon-20 contains five alpha particles. Most of the heavier atoms that are combinations of alpha particles need a few extra electrons in the nucleus in order to be stable. For example the uranium-236 nucleus is modeled as being comprised of 59 alpha particles with 26 extra electrons. All atomic nuclei except possibly iron-56 also contains gamma ray entrons that help hold the nuclei together and to provide some of their masses. When the nuclei of large atoms, such as uranium-235, split in fission reactions, some of these gamma ray entrons are released as fission energy, gamma rays. The construction of larger atoms from smaller atoms and particles takes place in stars and is explained in **Chapter XIII**.

Electricity and Magnetism

Everyone is familiar with electricity and magnetism, but the current scientific models do not provide a simple easy to understand explanation of either electricity or magnetism. I provide that explanation in **Chapter XIV**. Electrons are the source of both electricity and magnetism. Electricity is provided by energetic electrons which transmit electric energy in the form of entrons. Magnetism is provided by self-propelled naked electrons which travel south to north through magnetic material. These naked electrons exit at the north pole of the magnetic material and travel at a speed of about 2.19 X 10^6 m/s back to the south pole. These naked electrons have zero electrical energy but they each have a charge and with that charge they carry the Coulomb force. And this Coulomb force, produced by a stream of these naked electrons, can push energetic electrons back and forth in a copper wire to produce an alternating current in the copper wire.

Gravity and Anti-Gravity

Gravity is carried throughout our Universe by neutrino photons which are produced in Black Holes at the center of galaxies by the destruction of protons. Each proton destroyed releases one neutrino entron many of which travel out from the Black Holes as neutrino photons at the speed of light. Neutrino entrons are the smallest things in our Universe with the exception of tronnies that have no size at all. Most neutrino photons pass through stars, planets and moons applying a reverse force as they pass through. This reverse force is directed toward the source of the neutrino photons. A small percentage of the neutrino photons is temporarily stopped in stars, planets and moons and later released in random directions giving the stars, planets and moons their gravity. Low energy photons pass through interstellar space more efficiently than neutrino photons and are absorbed by or reflected from the matter of other galaxies producing a photon pressure on the other galaxies resulting in an anti-gravity, a repulsive force on the other galaxies. Neutrino photons apply a stronger attractive force on nearby galaxies but the anti-gravity effects of the low-energy photons wins out for the far-away galaxies. Therefore, far-away galaxies are receding from each other and close-by galaxies are attracting each other. Gravity and Black Holes are discussed in **Chapter XX**. Anti-gravity is discussed in **Chapter XXI**.

Coulomb Forces Hold Atomic Nuclei Together

Coulomb forces hold atomic nuclei together. There is **no "strong force"**. There is **no "weak force"**. The so-called "binding energy" is not some magical force that overcomes the repulsive forces of protons in the nuclei. Forces in the nuclei are only Coulomb forces positive and negative, all in balance. Gamma ray entrons are incorporated into each atomic nucleus as they are fabricated in stars and these gamma ray entrons can be released in processes such as radioactive decay and nuclear fission.

There are **no neutrons in the nuclei of stable atoms**. Neutrons decay with an average life of about 15 minutes whether they are inside nuclei or outside the nuclei. In the place of each neutron there is one proton, one electron and one gamma ray entron. Neutrons are discussed in **Chapter XII**.

During the first minutes, months or years after the Big Bang naked protons were formed from electrons, positrons and neutrino entrons. These naked protons collected gamma ray entrons to (1) slow down and (2) to cool off our early Universe. After the protons have slowed down sufficiently, they each capture an electron to become a hydrogen atom. Stars are currently converting these hydrogen atoms into larger atoms and releasing the gamma ray entrons to provide the heat and light for their orbiting planets and the things living on them. Heat and temperature are discussed in **Chapter XIX** and processes for building atoms are discussed in **Chapter XIII**.

The Shell of Our Universe

Our Universe is surrounded by **a cold plasma shell** of mostly high-speed electrons and positrons that reflect light and other lower energy radiation reaching the shell to provide the recently discovered **cosmic background radiation**. Many if not most of the gravity producing neutrino photons from the 100 to 400 billion galaxies in our Universe that make it to the shell are captured by electrons to become high-energy massive electrons, and these electrons can capture two positrons to form protons at the edge of our Universe. These protons slow down by capturing gamma ray entrons to become the nuclei of hydrogen atoms each of which in turn capture an electron to become hydrogen atoms to help provide the energy and matter for star formation at the outer regions of our Universe. The shell of our Universe is discussed in **Chapter XXII**.

Coulomb Grids

Everything in our Universe is comprised of tronnies. These tronnies are spread throughout our Universe in stars, planets, moons and interstellar space and each tronnie continually collects and produces Coulomb force waves that spread out from each tronnie at the speed of light. These speed-of-light force waves produce **Coulomb grids** in all solid matter, in liquids, in gasses and in interstellar space and everywhere else in our Universe. The speed-of-light waves within these grids can come to focus in points. These points of focus are the tronnies that our Universe is made of. So tronnies produce the Coulomb grids of our Universe and the Coulomb grids of our Universe produce the tronnies. Coulomb grids are discussed in **Chapter XXIII**. It is easy to imagine how tronnies could produce Coulomb grids. It is more difficult to imagine how Coulomb grids could produce tronnies. In **Chapter XXIV** I provide some details as to how tronnies are created from the Coulomb grids. *As a hint I will tell you now that each tronnie traveling in a circle, at π/2 times the speed of light, is always at the focus of its own Coulomb force waves traveling at the speed of light.*

As I am making the final edits to this book, many of the smartest scientists on earth are spending billions of dollars in search of the Higgs boson and the Higgs field. The Higgs field is supposed to be an invisible field that gives mass to everything that has mass. They are on a "wild goose chase". There is no Higgs boson. The "field" that gives mass to everything in our Universe that has mass is the Coulomb grids described in **Chapter XXIII**. Tronnies are points of focus of the speed of light Coulomb waves. Tronnies have no mass and no energy but they each have a charge and they each carry the Coulomb force and when they combine with other tronnies to make electrons and positrons, mass is produced. Two opposite tronnies produce an entron which has both mass and energy. It is these entrons (each with two tronnies) that provide all of the mass of our Universe other than the mass of electrons and positrons (each of which are made of three tronnies). And the single entron that provides most of the mass of our Universe is the neutrino entron. (I suppose some people might suggest that my neutrino photon is the Higgs boson.) It is not.

The Death of Our Predecessor Universe and the Creation of Our Universe

Our Universe is but one of a series of universes all of which have been born in a Big Bang explosion, expanded for a number of years, contracted for a number of years and died in a Big Bang explosion. Our predecessor universe was much like our Universe. About half-way through its life gravitational forces overcame anti-gravity forces and our predecessor universe began to contract. Eventually most of the Black Holes of our predecessor universe combined into one enormous last remaining Monster Black Hole. Neutrino photons from this last remaining black hole continued to drive the remains of our predecessor universe into the Monster Black Hole. Our predecessor universe died and our Universe was born in a Big Bang explosion about 13 to 15 billion years ago when neutrino photons from this Monster Black Hole of our predecessor universe had pushed almost all of the galaxies, of our predecessor universe including their stars, planets and moons, into this Monster Black Hole. This Big Bang explosion was the death of our predecessor universe and birth of our Universe. At the time of the Big Bang explosion, a small percentage, but a large number, of galaxies of our predecessor universe were barreling toward the Monster Black Hole at speeds thousands of times the speed of light! These tremendous speeds were the consequence of continuous gravational acceleration for many billions of years. When this large number of galaxies coming in from all directions arrived at the location of the Monster Black, it wasn't there! So these galaxies passed through the empty site of the Monster Black Hole and continued at substantially the same speed spreading out through our baby Universe in all directions at many thousand times the speed of light. **This was the inflation period of our Universe!** The Big Bang explosion released huge quantities of energy in the form of neutrino entrons, gamma ray entrons and lower energy entrons. These entrons spread out much faster than the speed of light along with the surviving galaxies as photons through the nearby regions and into the of the electron-proton shell remaining from our predecessor universe.

Entrons combined to produce electron-positron pairs in a process called "pair production". In this process neutrino entrons, gamma ray entrons and lower energy entrons combine to make the separate electrons and positrons. Each combination of the three entrons provides three plus tronnies and three minus tronnies. And the result of each combination is one electron-positron pair which together includes three plus tronnies and three minus tronnies.

Electrons, positrons and neutrino entrons then combine to produce high-speed naked protons. In this process of proton production an electron combines with the neutrino entron to produce a very high-speed, high-mass electron circling faster than the speed of light with a diameter of approximately 0.85×10^{-15} meters. The circling high-speed, high-mass electron captures two positrons to form a naked proton which is self-propelled at speeds of about thirteen percent of the speed of light. The naked proton captures several gamma ray entrons to slow down to speeds close to zero to become the low-speed, high-energy proton with which we are familiar. This capturing of gamma ray entrons by high-energy protons was a major factor in cooling off our early Universe.

The low-speed, high-energy proton captures an electron to become a hydrogen atom. Hydrogen atoms combine in enormous clouds, which are compressed by the passage of neutrino photons through the clouds. The compressed clouds of hydrogen atoms form stars and fusion reactions in stars convert hydrogen atoms into helium. In these processes four protons and two electrons combine to form an alpha particle. Some of the proton's gamma ray entrons are released as fusion energy. The alpha particle is also a high-speed particle and captures gamma ray entrons to slow down to become the nucleus of the helium atom. Each helium nucleus collects two electrons to form a helium atom. Helium nuclei, protons and electrons combine to form heaver nuclei of atoms up to iron with each combination releasing gamma ray entrons. Explosions of stars force additional combinations of atoms and gamma ray entrons to produce the atoms heavier than iron. These atoms from exploding stars spread about our Universe and later collect to form planets, moons and other things, eventually including us. Our Universe is currently expanding but ultimately it will start to contract and our Universe will ultimately be destroyed in a Big Bang explosion which will be the birth of our successor universe. Additional explanation of the life and death of universes is provided in **Chapter XXV**.

Where Do Tronnies Come From?

The Ross Model proposes that everything in our Universe is made from tronnies or things made from tronnies. But how do we make the tronnies? I provide answers in **Chapter XXIV.**

The Ross Atom

In **Chapter XXVI** I describe the hydrogen-1 atom, the smallest atom and the most important atom in our Universe. My description is similar to the Bohr atom described about 100 years ago by Niels Bohr.

Numbers to Remember

In **Chapter XXVI** I summarize some of the specific numbers needed to follow some of the calculations included in this book. These numbers may be of some help for those of you who want to check these calculations.

One Hundred and One Predictions

In **Chapter XXVII**, I list 101 predictions of the Ross Model. If you want, you may take a look at these predictions before you read any further and write "T" for the predictions you believe may be correct, "F" for the ones you think you disagree with and "?" for the ones with respect to which you don't want to take a guess. After you finish the book you may want to take the quiz again. If you can prove that any of the predictions are incorrect, I request that you get in touch with me. If you are the first to prove conclusively that some of my more controversial predictions are correct, you may be a candidate for the Nobel prize.

Mr. Ross believes he was the first person in our Universe to identify tronnies. Tronnies are point particles with no volume and no mass. But every tronnie has a charge of +e or -e so each of them produces the Coulomb force. According to the Ross Model, tronnies, in order to exist, must travel in perfect circles at speeds of $\pi/2$ times the speed of light. That is about 1.57 times the speed of light. Like charges repel and unlike charges attract. So each tronnie repels _itself_ faster than the speed of light!

Three tronnies (two minus and one plus) combine with each other to make an electron and three tronnies (two plus and one minus) combine to form a positron. Two oppositely charged tronnies combine to make an entron. Each electron and each positron has a fixed mass and volume and a charge respectively of minus e or plus e. Each entron has a diameter and a mass that varies by more than 16 orders of magnitude so the largest entrons are more than ten million billion times larger and less massive than the smallest entrons. Otherwise all entrons in our Universe are exactly alike. Entrons are described in Chapter IV and electrons and positrons are described in Chapter V. Electrons, positrons and entrons are composite building blocks of everything else in our Universe.

CHAPTER III

THE TRONNIE

No Mass, No Volume, Only Charge

The Ross Model is based on the existence of a previously unknown point particle which I first identified and described about eight years ago in a "thought experiment" and call the "tronnie". Tronnies have no mass and no volume but they do have a charge of plus e or minus e. (A charge of minus e is equivalent to the net charge of an electron and a charge of plus e is equivalent to the net charge of a positron, which is, as explained in **Chapter I**, the anti-particle of the electron.) By reason of their charges, tronnies are the source of the Coulomb force which expands out from each tronnie at the speed of light (about 3×10^8 m/s, i.e. 300 million meters per second) repelling like tronnies and attracting unlike tronnies. **Each tronnie, being exactly like itself, repels itself!** Tronnies travel in perfect circles with one or two other tronnies at speeds of $\pi c/2$ (about 1.57 times the speed of light).

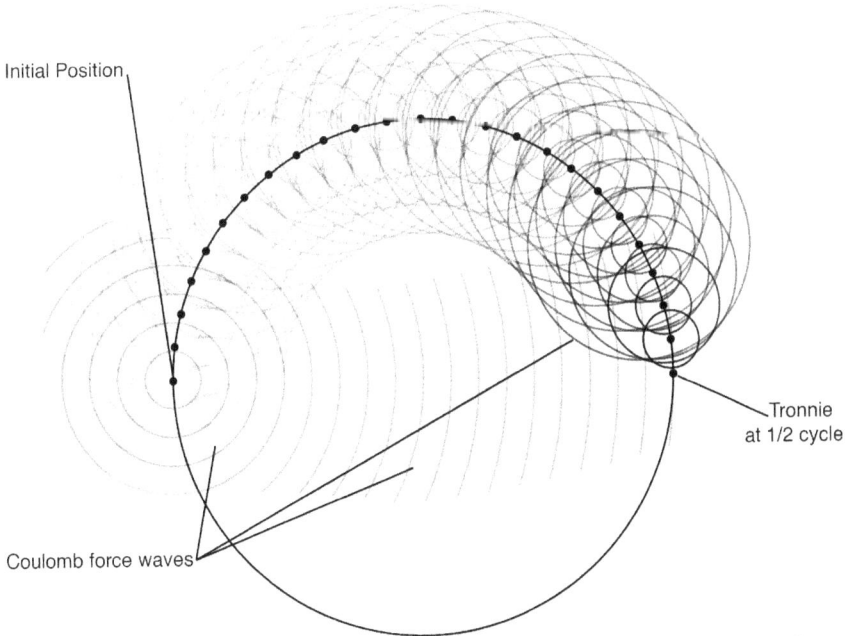

Initial Position

Tronnie
at 1/2 cycle

Coulomb force waves

FIG. 1

Tronnies are the basic building blocks of our Universe. Everything in our Universe is made from tronnies or things made from tronnies. There is nothing in our Universe that is not made entirely of tronnies.

By reason of its charge Coulomb force waves travel out from each tronnie continuously at the speed of light. But the tronnie (by traveling in a perfect circle at $\pi/2$ times the speed of light) is always located at a point which is the focus point of its own Coulomb force waves. **FIG. 1** shows the Coulomb force waves traveling out from a tronnie as the tronnie makes a trip half way around its circle from a point corresponding to a 180 degree position on the circle to a point corresponding to a zero degree position on the circle. Note that at the halfway point, the Coulomb force wave from its initial position at the 180 degree position of the circle has finally caught up to the tronnie at the zero degree position of the circle. At this point the tronnie is repelled in the zero degree direction by its own Coulomb force wave. But this relationship is true for each point on the tronnie's circle. At each point on the circle the tronnie is repelled by a Coulomb force wave coming diametrically across the circle. For example, when the tronnie gets back to the 180 degree position on the circle, it will be repelled in a 180 degree direction by a Coulomb force wave coming from the zero degree position on the circle.

FIG. 1A is a drawing of one half cycle of an entron which is comprised of one plus tronnie and one minus tronnie each traveling in the same circle at speeds of $\pi/2$ times the speed of light with each of the two tronnies continuously producing Coulomb force waves which travel out from the two tronnies at the speed of light. The plus tronnie is traveling from 180 degrees to zero degrees while the minus tronnie travels from zero degrees to 180 degrees. Each of the two tronnies are continuously repelling themselves and attracting the other tronnie. I will explain in **Chapter IV** how the attractive and repulsive forces on each tronnie in the diametrical direction are exactly equal while forces in the tangential direction keep the two tronnies circling continuously at about 1.57 times the speed of light.

Entrons, Electrons and Positrons are Made from Tronnies

A single tronnie combines with an oppositely charged tronnie in a circling pair to form a mass/energy quantum that I have named "entron". **FIG. 1A** is a drawing of an entron showing the Coulomb force waves of a plus tronnie traveling from the top of the drawing to the bottom and the Coulomb force waves of a minus tronnie traveling from the bottom of the drawing to the top. The paths of the two tronnies and positions of the two tronnies are shown for one half of a single cycle for the two tronnies. An entron traveling at the speed of light is a photon. Electrons are comprised of one plus tronnie and two minus tronnies, and positrons are comprised of one minus tronnie and two plus tronnies. I describe in detail entrons in **Chapter IV**, photons in **Chapter V**, and electrons and positrons in **Chapter VII**.

Everything Else is Made from Entrons, Electrons and Positrons

Entrons, electrons and positrons are made from tronnies and everything else in our Universe is made from entrons, electrons and positrons. Everything else includes photons, magnetism, gravity, electricity, protons, atoms, molecules, plants, animals, planets, moons, stars, galaxies and you and me, all made from tronnies or things (entrons, electrons and positrons) made from tronnies. Each photon is comprised of only one entron.

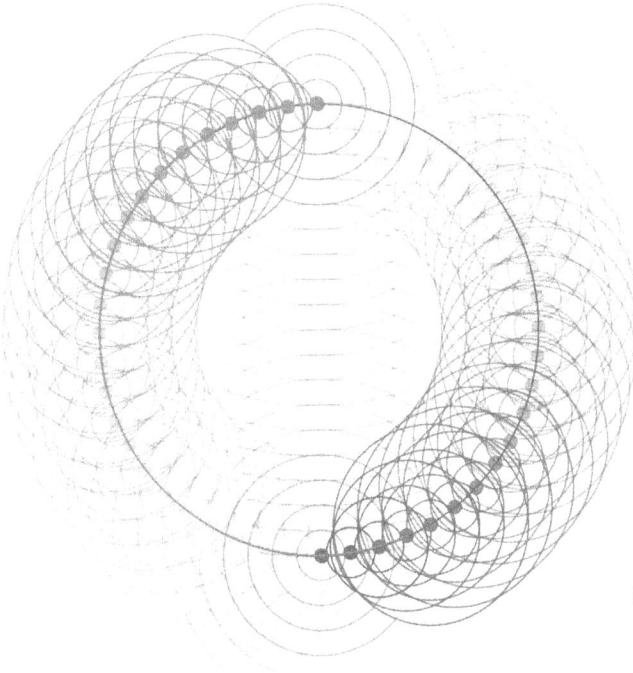

FIG. 1A

Tronnies Are Required by Law, Coulomb's Law

Coulomb's Law, as expressed in **Equation 2** in **Chapter I**, requires that all charged particles be point particles or comprised of point particles. **Equation 2** is:

$$F = kQ_1Q_2/r^2, \qquad (2)$$

29

As the distance r separating the particles approaches zero the force F approaches infinity. If a charged particle, comprised of a single charge, had any volume at all, Coulombs Law requires that the parts of the particle would repel other parts (with forces approaching infinity), so the particle would quickly blow itself apart. A particle comprised of a single charge therefore has to be a point particle. A point particle with a single charge is a **tronnie**. Electrons and positrons are very small, they each have a single **net** charge and they do not blow themselves apart. But electrons and positrons are **not comprised of a single charge**. As explained above, electrons and positrons are each comprised of three charges: two minus and one plus for the electron and two plus and one minus for the positron.

The concept of charge is a difficult concept. We know that particles with net charges, such as electrons, positrons and protons, as well as atomic ions and molecular ions produce a Coulomb force that spreads out at the speed of light. **Equation 2** describes the force between **static charges**. (The Ross Model provides a slightly different equation (**Equation 2A**) for the force between **rapidly circling charges**. That equation is:

$$F_I = \pi k Q_1 Q_2 / d' \qquad (2A)$$

where F_I is force "integrated" around the circle and d' is the diameter of the circle.) I will have more to say about this in **Chapter IV**.

The Size of the Tronnie

Let me make one more point here regarding the size of the tronnie. I explain that the tronnie has to be a point particle, with no size what-so-ever, to comply with Coulomb's Law. How it can be a point particle is explained in more detail in **Chapter XXII**. The tronnie's point is a point of focus of Coulomb force waves and those Coulomb force waves expand out in waves from the tronnie at the speed of light. These waves expand forever although they become very weak very quickly. Nevertheless, they do expand out forever. Therefore, you could think of the size of the tronnies as including its expanding Coulomb waves. In that case each tronnie can have any size you choose.

The Tronnie's Charge

Existing scientific models explain that charge is a physical property of a particle that causes it to experience a force when near other particles having a charge. Charge is either positive or negative and the force resulting from the charge is the Coulomb force that spreads out in all directions from particles with charge at the speed of light. Charge is quantized, that is, it comes in individual small units called the elementary charge, e, approximately equal to 1.602×10^{-19} coulombs, all as described in **Chapter I**. Existing models do not have a good explanation of where charge comes from. According to the

Ross Model the charge of all charged particles in our Universe is the net charge of all of the tronnies that the particles are comprised of. Now, the question is, "Where do tronnies get their charge?"

This issue is dealt with in detail in **Chapter XXIV**. Each tronnie gets its charge from the Coulomb grid in which it is located or through which it is passing. Coulomb grids, as described above, are produced by tronnies and nothing but tronnies. As explained in **Chapter XXIV** much of, if not all, the charge is of each tronnie is produced by the tronnie itself traveling in a circle at a speed of $\pi c/2$, so the tronnie is always located at the focus of its own Coulomb force waves traveling diametrically across the tronnie circle. In **FIGS. 1 and 1A** the two tronnies are viewed from a distant location perpendicular to the plane to the circle of the tronnies so that the path of the tronnies around their circle can be shown.

Coulomb force waves from each of the two tronnies in each entron are <u>focused</u> on each of the two tronnies. As the Coulomb forces come to a focus at the point which is the tronnie, they spread out in 360 degrees to give the point (i.e. the tronnie) its charge. As the force waves of these tronnies pass through the point (which is and represents our tronnie) they spread out in all direction giving the tronnie its charge. If you are confused about the origin of charge, don't worry about it right now, I will have much more to say about this complicated concept later on.

Coulomb Grids

Every tronnie in our Universe is exactly like every other tronnie except one half of them have a charge of plus e and the other half have a charge of minus e. These tronnies are present everywhere in our Universe, but being point particles they occupy no space. All are traveling at $\pi c/2$ in at least one reference frame. Every one of these tronnies is producing Coulomb force waves that are traveling out from the tronnies at the speed of light producing Coulomb grids. So we have a tremendous number of Coulomb grids including a universe Coulomb grid that is stationary with respect to our Universe. All other Coulomb grids travel through the universe Coulomb grid at a great variety of velocities. In **Chapter XXIII** I will have more to say about Coulomb grids.

The entron is the basic energy-mass quantum in our Universe. The diameter of entrons can be any distance from about 0.934×10^{-18} meters to a few centimeters, the smaller the diameter the larger the energy-mass of the entron. As far as we know, like the tronnies, Mr. Ross was the first person in our Universe to recognize the existence of the entrons.

As indicated in FIG. 2A, with two opposite tronnies traveling in a perfect circle at speeds of $(\pi/2)c$, the attractive integrated forces between the two tronnies on each other in the diametrical direction is exactly equal to the repulsive integrated forces of each tronnie on itself from across the diameter of the entron circle. Proof is provided in Chapter VI. So once created by the combination of two opposite tronnies, an entron can exist for billions of years.

Mass is a property of a thing that allows the thing to resist an outside force. The forces acting on the tronnies and integrated around the entron circle combine to allow the entron to resist any outside force. So, two particles with opposite charges but no mass combine to make a particle with mass but no net charge. The entron also has energy corresponding to its mass according to Albert Einstein's famous formula: "$E = mc^2$".

CHAPTER IV

THE ENTRON

A Quantum of Energy

Entrons are extremely important; they represent all of the thermal energy, all of the radiant energy, all of the chemical energy, all of the nuclear energy of our Universe and probably all of the other forms of energy in our Universe. This may include even kinetic energy except for the kinetic energy associated with the natural speed of electrons and positrons. Entrons are also a quantum of mass. Entrons represent almost all of the mass of our Universe. The only other basic particles with mass are electrons and positrons (each pair of which is the result of a combination of three entrons). Every entron is comprised of two tronnies, a plus tronnie and a minus tronnie. The two tronnies of each entron travel on opposite sides of a circle at a speed of exactly $\pi/2$ times the speed of light (which is about 1.5708 c). All entrons are exactly alike except for the size of their circles which determines their energy and mass. The diameter of the circle may be any size from 0.934 X 10^{-18} m (the neutrino entron) to about 1 X 10^{-1} m, about 10 centimeters (low-energy radio wave entrons). (To put things in prospective, the size (diameter) of atoms is in the range of about 10^{-10} m and the size (diameter) of atomic nuclei is roughly about 10^{-15} m. The size of the smallest entrons (that I have been able to identify) at 0.934 X 10^{-18} m, is about one half the size of the electron and about a million times smaller than a typical atom. The largest entrons are about 100 million billion times larger than the smallest entron. And the mass and energy of the smallest entrons are about 100 million billion times larger than the mass and energy of the largest entrons. Otherwise all entrons are exactly alike and are as pictured in **FIG. 2A**.

The frequency of an entron is the number of cycles per second made by the two tronnies. The frequency f is:

$$f = c/2d'$$

which is the speed of light (c, about 3 X 10^8 m/s) divided by twice the entron diameter, d'; so the entron frequencies range from about 150 trillion, trillion cycles per second (150 X 10^{24}/s) to about 1.5 billion cycles per second (1.5 X 10^9/s).

I suggest that the reader stop for a few minutes and think about the frequencies at which these tronnies circle in these entrons. They make their trips around the circumference of the entron circle at a rate of about 1.5 billion times per second for the lowest energy entrons to about 150 trillion trillion times per second for the highest energy entron.

33

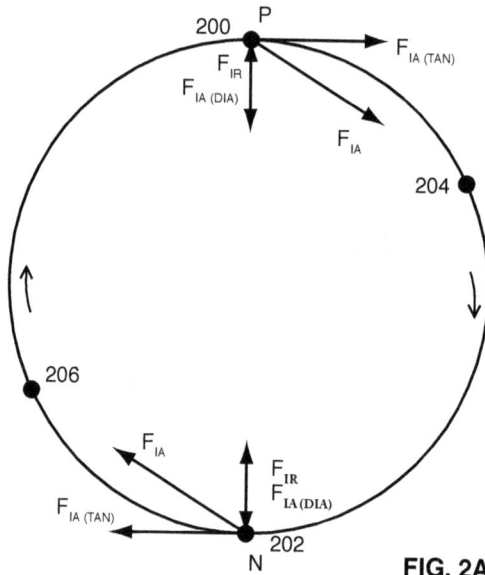

FIG. 2A

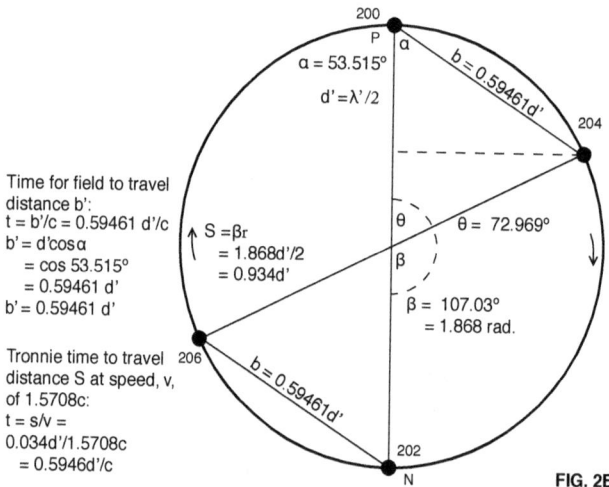

Time for field to travel distance b':
$t = b'/c = 0.59461\ d'/c$
$b' = d'\cos\alpha$
$\quad = \cos 53.515°$
$\quad = 0.59461\ d'$
$b' = 0.59461\ d'$

$\alpha = 53.515°$
$d' = \lambda'/2$
$b = 0.59461d'$

$S = \beta r$
$\quad = 1.868d'/2$
$\quad = 0.934d'$

$\theta = 72.969°$

$\beta = 107.03°$
$\quad = 1.868\ rad.$

Tronnie time to travel distance S at speed, v, of 1.5708c:
$t = s/v =$
$0.034d'/1.5708c$
$\quad = 0.5946d'/c$

$b = 0.59461d'$

FIG. 2B

During each trip each tronnie is continuously intersected by its own Coulomb force wave coming from all directions. Thus, each tronnie becomes a "focus point" of its own

Coulomb force waves coming from all directions 1.5 billion to 150 trillion trillion times per second. (You may wonder what happens to these Coulomb force waves after they pass through the point of focus which is the tronnie.) This may provide a hint as to where the tronnie gets its "**charge**". We will have more to say about charge in **Chapter XXIV**.

As explained above a single entron is the energy part of every photon. Entrons provide the heat and the temperature of everything in our Universe. One of these entrons (the neutrino entron) along with one electron and two positrons are the building blocks of protons. Protons, electrons and entrons are the building blocks of atoms, and atoms in turn are the building blocks of everything else in our Universe. Except for a relatively very small mass contribution from electrons and positrons, entrons supply all of the mass of our Universe including the dark matter of our Universe. The "neutrino entron" (which we will examine in detail in **Chapter VI**) provides almost all of the mass of the proton and carries the gravity of our Universe. Entrons like tronnies were discovered by me in a thought experiment about eight years before the publication of this book. Please refer to **FIG. 2B**, there you will see a snap-shot drawing of an entron showing the precise location of minus tronnie (N) and plus tronnie (P) at the time of the snapshot. The Coulomb force waves of each of the two tronnies travel diametrically, a distance d', across the entron circle at the speed of light c; while the tronnies themselves take a d'π/2 longer half-circle route on the circumference of the circle at a speed of πc/2, so each of the two tronnies are continuously meeting their own Coulomb force wave on the opposite side of the circle. So each of the two tronnies are continuously being repelled by their own Coulomb force waves with a force of F_{IR} (short for "integrated repulsive force") coming diametrically from the opposite side of the circle (e.g. from point 202 to point 200 and from point 200 to point 202 as shown in **FIG. 2A**). Each of the two tronnies are also being attracted by Coulomb attractive forces, F_{IA} (short for "integrated attractive force") from their partners, but these attractive force waves travel a shorter distance (e.g. from point 204 to point 200 and from point 206 to 202); so the F_{IA} attractive forces are stronger forces than the F_{IR} repulsive forces as indicated by the lengths of the arrows shown in FIG. 2A. These integrated Coulomb forces are inversely proportional to the distance (not the square of the distance) the forces have to travel. The equation for the integrated repulsive forces F_{IR} is:

$$F_{IR} = \pi k Q_1 Q_2 / d' \qquad (3A)$$

where k is the Coulomb constant = 8.99 X 10^9 Nm^2/C^2, Q_1 and Q_2 are the charges of the tronnies, each equal to the electron charge of about 1.602 X 10^{-19} coulomb, and d' is the diameter of the entron's circle. The equation for the integrated attractive forces is:

$$F_{IA} = \pi k Q_1 Q_2 / b' \qquad (3B)$$

But the integrated attractive force in the diametrical direction is:

$$F_{IADIA} = F_{IA}\cos\alpha = (\pi k Q_1 Q_2/b')\cos\alpha = F_{IR} \quad (3C)$$

since $\cos\alpha = b'/d'$ as indicated in FIGS. 2A and 2B.

Equations (3A and 3B) will be further developed in **Chapter VI** where we will consider the forces within the neutrino entron.

Attractive and Repulsive Forces are Equal

The secret of the entron's success is that the component of the attractive integrated force F_{IA} in the diametrical direction $F_{IA(DIA)}$ is exactly equal to the repulsive integrated force F_{IR} which is also in the diametrical direction as indicated in **FIG. 2A** and **Equations 3A** and **3C**. So the two forces in the diametrical direction cancel each other. The component of the attractive integrated forces in the tangential direction $F_{IA(TAN)}$ is not cancelled and this force assures that the two tronnies will continue circling along their circumferential path.

FIG. 2B provides some hints for readers who may want to check my trigonometry. The time required for tronnie N to travel from point 206 to point 200 is exactly equal to the time for its Coulomb force wave to travel from point 206 to point 202. The energies and masses of the entrons are inversely proportional to their diameters d' which, as explained above, can be any length from about 0.934×10^{-18} m to at least few centimeters. The angle α in FIG. 2B is about 53.515 degrees, but people good at trigonometry will recognize that as long as the angle α is between zero degrees and 90 degrees the attractive and repulsive forces in the diametrical direction will be exactly equal. I consider this an amazing discovery by me! Others may think it obvious but so far as I know, no one has used this result to describe the basic energy quantum of our Universe.

No Need for the Higgs Boson

These simple entrons (each comprised of two tronnies – one plus and one minus) are the particles that give mass to other particles. As I write this book hundreds of scientists are on a wild goose chase spending billions of dollars looking for the Higgs boson which is supposed to give mass to other particles. There is no such thing as the Higgs boson! Entrons supply all of the mass of our Universe except for the small fraction that is supplied by electrons and positrons.

The Amazing Entron

Although tronnies are point charges with no size, no mass and no energy, when a single tronnie combines with another oppositely charged tronnie to form an entron, all of a sudden the pair has a size, energy and mass. The entron is truly an amazing particle that remained hidden from scientists for hundreds of years until now! Trapped in matter entrons represent thermal energy, nuclear energy, chemical energy and electrical energy. They are released from matter as photons which travel through space at the speed of light. (As we explain in **Chapter V**, entrons travel within the photon in a circle at twice the speed of light; thereby defining the photon's frequency.) The energy of the entron is the energy of the released photon which corresponds to the photon wavelength according to **Equation 1**.

$$E_{photon} = E_{entron} = hc/\lambda \quad (1)$$

The mass of the entron is the same as the mass of the photon which is determined from Einstein's famous equation: $E = mc^2$ or $m = E/c^2$, so the mass of the entron and the mass of the photon is determined by the photon wavelength λ by:

$$m_e = E/c^2 = hc/\lambda c^2 = h/\lambda c = 2.2087 \times 10^{-42} \, kg\text{-}m/\lambda$$

For example the wavelength of a typical green light photon is 5.42×10^{-7} m, so its mass is about 4.80×10^{-36} kg which is an extremely small mass. However, a 1.02 MeV gamma ray photon has a mass equal to the mass of two electrons (i.e. about 1.82×10^{-30} kg). This extension of Einstein's famous equation is very important since it demonstrates that photons have a mass which is inconsistent with the commonly held belief that anything with mass cannot travel at the speed of light. Examples covering the full range of the electromagnetic spectrum are provided in **Table V** in the next chapter. The reader should notice that the masses of all photons except the most energetic photons are very small and these masses typically go un-detected. But all photons, contrary to popular beliefs, have some mass that is inversely proportional to their wavelengths as indicated by the above equation.

The pair of tronnies (in the entron and the photon) has no net charge. We will see in **Chapter VII** that three tronnies (one plus and two minus) make an electron and three tronnies (one minus and two plus) make a positron. My recognition and reporting of the existence of the **tronnie** and the **entron** and the internal structure of **positrons and electrons** allows us to understand how entire universes with billions of galaxies can be constructed from a single type of particle and its anti-particle, each having no mass and no volume. **I believe that this recognition may be understood someday, relatively soon, as one of the most important scientific development in the past one hundred years!**

Photons are entrons traveling in a circle at a speed of twice the speed of light and forward at a speed of light. Each of the two tronnies of the entron wants to remain in its circular configuration and at the same time stay ahead of its own Coulomb forces coming from its earlier positions in the entron. To do this each entron needs to travel at a speed of 2c, but at a speed of 2c the entron would, if it traveled in a straight line, quickly outrun its own Coulomb force waves, which are traveling at a speed of c.

So the entron has its cake and also eats it. It travels in a circle at a speed of 2c and forward at a speed of c. This is a photon. FIGS. 2C and 2D show the two tronnies leapfrogging each other as the entron travels at a speed of 2c in a portion of a circle. FIG. 3 shows what a photon would look like if you were a small superman and flying along beside it a speed of c. FIG. 4 shows what the photon would look like if you were merely watching it pass by. The ratio of the photon diameter to the entron diameter is: $d/d' = 911$.

Table V lists some examples of typical photons and the entrons that provide the energy and mass of the photons, from the smallest to the largest photons. Note that the mass of the most massive photon (the neutrino photon) is almost equal to the mass of a proton.

38

CHAPTER V

PHOTONS

An Entron Traveling in a Circle and a Straight Line

According to the Ross Model each photon is an entron traveling in a circle at a speed of 2c (twice the speed of light!) and forward at a speed of c (the speed of light). **FIGS. 2C** and **2D** show 2¼ cycles of an entron traveling as a photon. In the entron frame of reference (moving at a speed of 2c) each of the two tronnies in the entron are pushed by a Coulomb force coming (at a speed c) diametrically across the entron circle from itself and is being attracted by attractive stronger forces from its partner. However the diametrical component of that attractive stronger force is exactly equal to the diametrical repulsive force from itself as explained in **Chapter IV** with reference to **FIGS. 2A** and **2B**.) In the entron frame of reference its circle is shown as 199 in FIGS. 2C and 2D. In the photon frame of reference Coulomb force waves are shown as 104, 106, 108 and 110 in **FIG. 2D**. In **FIG. 2D** only four Coulomb force waves are shown expanding from four points. The reader should note that the force waves are created continuously as the entron move through space.

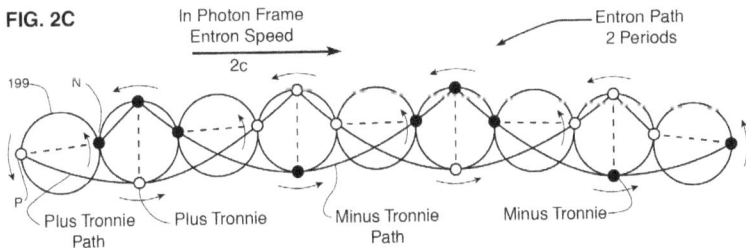

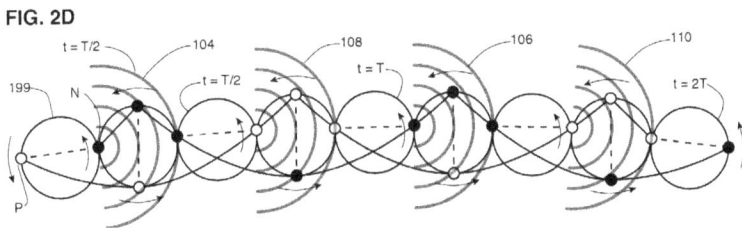

FIG. 2C

In Photon Frame
Entron Speed
2c

Entron Path
2 Periods

199   N

P

Plus Tronnie Path   Plus Tronnie   Minus Tronnie Path   Minus Tronnie

FIG. 2D

t = T/2   104   108   106   110

199   N   t = T/2   t = T   t = 2T

P

The diameter d of the photon's circle is related to the photon's wavelength λ by the following formulas:

$$d = \text{about } 0.6366\,\lambda \qquad (4)$$
$$\lambda = \text{about } 1.547d$$

This relationship result from the fact that the time for the entron to travel around the circumference of its photon's circle, a distance of πd (remember π = about 3.1416 and the circumference of a circle is πd) at a speed of $2c$, is the same as the time for the entron to travel forward a distance, equal to the photon wavelength λ, at a speed of c. Time t = distance divided by speed, so:

$$t = \pi d/2c = \lambda/c,$$

therefore the diameter d of the photon is:

$$d = 2\lambda/\pi = 0.6366\lambda$$

FIG. 3

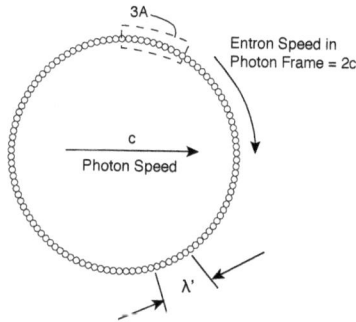

3A

Entron Speed in
Photon Frame = 2c

c
Photon Speed

λ'

FIG. 4

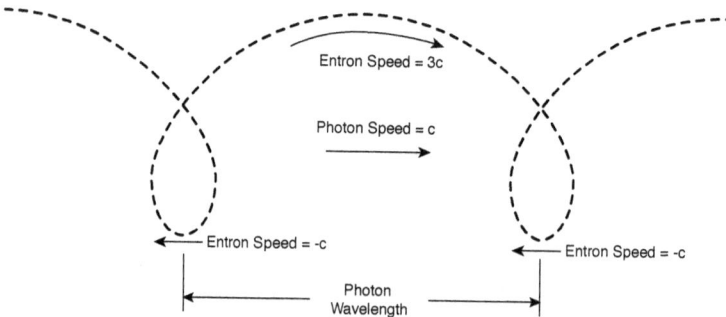

Entron Speed = 3c

Photon Speed = c

Entron Speed = -c

Entron Speed = -c

Photon
Wavelength

The Ross Model proposes that there is also a precise relationship between the size of the entron circle as shown in **FIG. 2A** and **2B** and the size of its corresponding photon circle as shown in **FIG. 3**. (The size of the entron circles is exaggerated in **FIG. 3**.) In **Chapter XI** I have estimated that the ratio of the diameter of the photon's circle d to the diameter of the entron's circle d' is approximately 911, i.e.:

d/d' = 911.

As indicated by **FIGS. 2C** and **2D** the entron makes a complete cycle in a distance equal to four times its diameter in its looping photon path. This would mean (if all of my assumptions are correct) that the entron makes about 1,431d'/4d' = 357.75 cycles in one photon circle and one photon wavelength. So:

$d/d' = 911$ (5)
$\lambda/d' = 1,431$ (6)
entron cycles/photon period = about 357.75 (7)

The reader should recognize the importance of these ratios. The ratio d/d' is a comparison of the size of a photon's circle (see **FIG. 3**) with the size of the entron's circle (see **FIG. 2A** in **Chapter IV**). There is substantial support in experimental evidence of my estimate of about 1,431 for the ratio of λ (the photon's wavelength) to d' (the entron's diameter). We know that gamma rays (with wavelengths λ in the range of about 10^{-12} m) are released from the nuclei of atoms which have sizes (about 1000 times smaller) in the range of about 10^{-15} m We also know that visible light photons with wavelengths of about 10^{-7} m are released from atoms which have sizes of about 10^{-10} m (about 1000 times smaller). I am hopeful of confirming my current estimate of 1,441 as the correct ratio of λ/d' (or at least that it is close to correct) but after about three years of worrying with this issue, I have not yet confirmed it. I will continue to use 1,441 until I or someone else develops a better number. If a better number is found, it will be easy to make the appropriate adjustments to the Ross Model.)

Once entrons which have been trapped in matter are released from the matter the Coulomb forces expanding out from the two tronnies, making up the entron, force the entron into the complicated path shown in **FIG. 4**. The entron itself keeps its circular configuration. However, both tronnies in the entron are attempting to stay ahead of their own Coulomb force waves as the tronnies are leapfrogging each other as shown in **FIGS. 2C** and **2D**. This results in the entron being propelled at twice the speed of light (2c = about 6 X 10^8 m/s) as shown in **FIGS. 2C and 2D**. The entron does not travel in a straight line at a speed of 2c. Since the Coulomb force waves from each of the two tronnies are traveling at a speed of c (c = 3 X 10^8 m/s), over any distance that is large compared to the diameter of the entron, the entron wants to travel at a speed of c, otherwise it would quickly outrun most of its own Coulomb force waves. How can the entron travel at a speed of 2c and also travel at a speed of c? Easy, it just travels in a circle at a speed of 2c as shown in **FIG. 3** defining a photon circle and the circle travels at a speed of c. The net effect is shown in **FIG. 4**.

Table V
Typical Photons

Photon	Photon-Entron Energy (eV)	Photon Wavelength (m)	Entron Diameter (m)	Black Body Peak Temperature (K)	Mass (kg)
Radio Wave Photons	1×10^{-8}	1.24	8.67×10^{-2}	2.34×10^{-4}	1.78×10^{-42}
H 21 cm Line	5.875×10^{-6}	2.11×10^{-1}	1.47×10^{-4}	1.37×10^{-2}	1.04×10^{-41}
Microwave Photons	1.02×10^{-5}	1.21×10^{-1}	8.5×10^{-5}	2.4×10^{-2}	1.83×10^{-41}
CBR (Peak)	2.35×10^{-4}	5.28×10^{-3}	3.69×10^{-6}	5.49×10^{-1}	4.18×10^{-40}
Millimeter Wave Photons	3.07×10^{-4}	4.04×10^{-3}	2.8×10^{-6}	7.17×10^{-1}	5.52×10^{-38}
Near Infrared Photon	0.124	1.0×10^{-5}	6.94×10^{-9}	290	2.21×10^{-37}
Std. Amb. Temp. 25C	0.125	9.92×10^{-6}	6.93×10^{-9}	293	2.98×10^{-37}
Human Body Temperature	0.133	9.32×10^{-6}	6.51×10^{-9}	310	2.37×10^{-37}
Visible Light Photons – Red	1.88	6.60×10^{-7}	4.61×10^{-10}	4,391	3.35×10^{-36}
Visible Light Photons - Green	2.29	5.42×10^{-7}	3.77×10^{-10}	5,350	4.08×10^{-36}
Sun Surface Temperature	2.48	5.00×10^{-7}	3.50×10^{-10}	5,800	4.42×10^{-36}
Visible Light Photons – Blue	2.7	4.59×10^{-7}	3.21×10^{-10}	6,300	4.80×10^{-36}
5 eV Ultraviolet Photon	5.00	2.48×10^{-7}	1.7×10^{-10}	1.16×10^{4}	8.90×10^{-36}
10.2043 eV Photon	10.2043	1.21×10^{-7}	8.49×10^{-11}	2.38×10^{4}	1.82×10^{-35}
13.6057 eV Photons	13.6057	9.14×10^{-8}	6.38×10^{-11}	3.17×10^{4}	2.43×10^{-35}
Extreme Ultraviolet	1.122×10^{3}	1.104×10^{-9}	7.722×10^{-13}	2.61×10^{6}	1.99×10^{-33}
X-Ray Photons	12.4×10^{3}	1.00×10^{-10}	6.99×10^{-14}	2.90×10^{7}	2.21×10^{-32}
Gamma Ray Photon	7.47×10^{5}	1.66×10^{-12}	1.16×10^{-15}	1.73×10^{9}	1.33×10^{-30}
Pair Production Gamma Ray	1.0214×10^{6}	1.213×10^{-12}	8.485×10^{-16}	2.38×10^{9}	1.82×10^{-30}
Gamma Ray Photon	8.37×10^{6}	1.48×10^{-13}	1.03×10^{-16}	1.95×10^{10}	1.49×10^{-29}
Neutrino Photon	9.28×10^{8}	1.335×10^{-15}	0.9339×10^{-18}	2.16×10^{12}	1.65×10^{-27}

Albert Einstein in developing his special theory of relativity wondered what a photon would look like if he could have caught up with it and observed it. If he had followed up on his question he might have concluded that the photon in its own frame of reference would look like the one pictured in **FIG. 3**.

With the entron circling at a speed of two times the speed of light and proceeding in a forward direction at the speed of light (c = 3 X 10^8 m/s); the entron path looks like the drawing in **FIG. 4**. This path looks somewhat like the path that a small light would describe if it were mounted on the rim of a wagon wheel rolling across the prairie at night. The perfect analogy would be a light on the rim of circular frame where the light on the rim is traveling at twice the speed of the wagon instead of only 1.57 times the speed of the wagon which is the normal speed of a point on the rim of a wheel. Remember, the two tronnies of the entron are continuing to travel in their own circle which has a diameter that is about 911 times smaller than the diameter of the photon circle.

From the view of a stationary person watching the photon pass by, the entron actually travels backwards at a speed equal to the speed of light at one point during each cycle of the photon as shown in **FIG. 4**. Its fastest speed is 3c and its average speed in the direction of the photon travel is c, all as shown in **FIG. 4**.

The reader should take a few minutes to study **Table V**. These 21 example photons represent particular photons of the electromagnetic spectrum that ranges in energy eighteen orders of magnitude from about 10 billionths of an electron volt to almost one billion electronvolts and includes: radio wave photons, microwave photons, millimeter wave photons, infrared photons, visible light photons, ultraviolet photons, x-ray photons, gamma ray photons and neutrino photons.

When entrons radiate away from matter in the form of photons the wavelength and the diameter of each photon are determined by the energy of the photon which is the same as the energy of its entron. This relationship between photon energy E_p and wavelength λ has been known for more than 100 years and is described by **Equation (1)** from **Chapter I** (i.e. $E_p = hc/\lambda$). Substituting the value for hc from **Table IV**, the photon energy is:

$$E_p = 19.86447461 \times 10^{-26} \text{ Jm}/\lambda \qquad (7)$$
$$E_p = 12.38942435 \times 10^{-7} \text{ eVm}/\lambda \qquad (8)$$

where eV is electron volts, J is joules and m is meters.
Also:

$$E_p = 7.8938 \times 10^{-7} \text{ eVm}/d$$
$$E_p = 8.6665 \times 10^{-10} \text{ eVm}/d'$$
$$E_p = 4.133 \times 10^{-15} \text{ eVm}/f$$

And:

$$E_p = 1.39 \times 10^{-28} \text{ Jm}/d' \text{ and}$$
$$d' = 8.6665 \times 10^{-10} \text{ eVm}/E_p$$

The energy of the photon E_p is the same as the energy of its entron E_e, so the energy of the entron E_e is also:

$$E_e = 1.39 \times 10^{-28} \text{ Jm}/d'$$

The frequency of the photon v_p is related to the frequency of the entron v_e by:

$$v_p = v_e/357.75$$

When photons are absorbed in matter (such as in the body of a pretty young girl lying on the beach soaking up the photons from the sun) the entrons that were the energy and mass of the photons then become integral parts of her body. She is absorbing visible, ultraviolet and infrared photons and her warm body is radiating mostly infrared photons and millimeter wave photons. Most of the photons absorbed by her body are absorbed as heat energy warming up her body. Photons radiating out from her body carry heat energy out of her body. Much of the visible light from the sun reflects off her body and bathing suit allowing people to see her and her bathing suit. Entrons in matter may exist in the form of heat energy increasing the temperature of the matter or it could be absorbed chemically and become a part of a new molecule in the matter. For example some of the ultraviolet entrons absorbed in her skin is used by her skin cells to produce vitamin D. Entrons of sunlight photons absorbed in leaves of plants are used (along with carbon dioxide in the air, water, atoms and molecules from the soil) by the plants to produce organic molecules allowing the plant to grow. Each of those entrons represents a small amount of energy. When a person consumes portions of the plant directly (such as when she eats a lettuce leaf or a carrot) or indirectly (such as when she drinks a glass of milk), those entrons become available to provide energy for her legs to propel her across a tennis court or a soccer field. When you burn a log in a campfire you are releasing visible light entrons that the leaves of a tree captured from photons radiated from the surface of our sun. No green light comes from the campfire since most of the green light photons were reflected by the tree leaves.

TABLE VI
GREEN LIGHT ENTRONS AND GREEN LIGHT PHOTONS

	Green Light Entrons	Green Light Photons
Diameter	3.77×10^{-10} m	3.43×10^{-7} m
Mass	4.08×10^{-36} kg	4.08×10^{-36} kg
Temperature to Produce	5,350 K	5,350 K
Entron Speed within Photon	2c (6×10^8 m/s)	
Photon Speed		c (3×10^8 m/s)
Energy	2.29 eV	2.29 eV
Wavelength		5.4×10^{-7} m

Table V is a comparison of features of green light entrons to features of green light photons.

Photon Mass
According to most current science texts, photons have no mass which allows them to travel at the speed of light. The Ross Model proposes that all photons have mass and that

their mass is determined by the photon's energy which is determined by Equation (1):

$$E = hc/\lambda$$

The mass of the photon and the mass of its entron is determined by Albert Einstein's famous equation:

$$E = mc^2$$

$$m = E/c^2 = (hc/\lambda)/c^2 = h/\lambda c = 6.626 \times 10^{-34} \, Js/(3 \times 10^8 \, m/s)\lambda$$

$$m = 2.2087 \times 10^{-42} \, kg\text{-}m/\lambda$$

$$m = 1.78 \times 10^{-36} \, kg/eV$$

Notice from **Table III** in **Chapter I** that the mass of the electron (9.109×10^{-31} kg) is about two hundred thousand times more massive than the green light photon. This may be a good reason why scientists have believed for many years that photons have no mass. They may have also been confused by Professor Einstein's theories that suggested that anything with mass could not travel at the speed of light. Under the Ross Model all photons have mass, even radio wave photons which are about a 50 million times less massive than the green light photon. On the other hand as indicated in **Table V**, the 1.02-MeV gamma ray photon is twice as massive as an electron and the neutrino entron and the neutrino photon are about four hundred million times more massive than the electron!

The Doppler Effect and Special Relativity

We know that the **measured** vacuum speed of light (as confirmed by Michelson & Morley about 100 years ago) is always constant at about 3×10^8 m/s. We also know that if we are moving toward a light source, or the light source is moving toward us, the wavelength of the light we measure will be reduced and the frequency and energy of the photons will be increased. Albert Einstein dealt with these facts with his special theory of relativity. He incorrectly interpreted the M&M results to mean the **actual** speed of light is always constant at about 3×10^8 m/s. The Ross Model provides a much simpler explanation. If a beam of light is entering a Coulomb field which is moving in a direction opposite the direction of the beam, the speed of the beam will decrease so that the photons continue to travel at the speed of light through the Coulomb field. In order to slow down the diameter of the photon circle (see **FIG. 3**) shrinks and its frequency and energy increases. This is the "Doppler Effect". The opposite is true if the Coulomb field is moving in the same direction of the beam of light. Our earth has a Coulomb field that moves with our earth as our earth moves through the solar system and our solar system has a Coulomb field that moves with the solar system as the solar system moves through the Milky Way Galaxy and the Coulomb field of the Milky Way Galaxy moves with the Milky Way galaxy as the Milky Way Galaxy moves through the Coulomb field of our Universe. Light from distance stars and galaxies must continually change speeds so as to travel at a constant speed in the various Coulomb fields on its way to our earth and within the Earth's Coulomb field. We measure that constant speed at about 3×10^8 m/s.

Neutrino entrons were discovered by Mr. Ross several years ago. Except for tronnies they are the most important character in his Ross Model of our Universe.

One neutrino entron provides more than 99 percent of the mass of each proton. One of these neutrino entrons combines with a gamma ray entron and another entron (six tronnies) to produce an electron and a positron pair (six tronnies) in pair production processes.

Neutrino entrons are produced in black holes at the center of each galaxy with the destruction of protons and exit the black hole at the speed of light as neutrino photons. As they pass through things, the neutrino photons produce a reverse Coulomb force on charges in the things to provide the gravity that holds galaxies together. Some neutrino entrons are captured temporally in stars, planets and moons; then they are released later in random directions as neutrino photons to give these objects their gravity. Estimates of 34,000 of these neutrino photons from the Black Hole in the center of our Galaxy are penetrating your body each second. Additional neutrino photons from the earth, our moon and our sun are also penetrating your body each second. You will see a calculation supporting this estimate in Chapter XX.

The neutrino photon flux exiting Black Holes is so large that even light is forced back toward the Black Holes by the reverse Coulomb forces of the neutrino photons.

CHAPTER VI

NEUTRINO ENTRONS AND NEUTRINO PHOTONS

The Most Energetic

The most energetic entrons and the most important entrons in our Universe are the "neutrino entrons" (each with a diameter d'_{ne} which I estimate to be about 0.934×10^{-18} m, a diameter that is about 100 million times smaller than the size of a typical atom). Its diameter is equal to the diameter of the paths of each of the three tronnies in the electron and the positron which are described in **Chapter VII**. Its corresponding photon, the neutrino photon, is the most important photon in our Universe. Its wavelength λ_{np} is about 1.336×10^{-15} m and the circle of this photon has a diameter of about 0.85×10^{-15} m, (equal to the diameter of the circular path of the high-energy electron in the proton) about one-half the size of a proton, the main component of the nucleus of a hydrogen atom. The neutrino entron and the neutrino photon each have a mass of about 1.6546×10^{-27} kg and an energy of about 1.487×10^{-10} J (or about 928×10^{6} eV). I will explain how I have estimated these values in **Chapter X**. I am the first person in our Universe to recognize the existence of neutrino entrons and neutrino photons. I have documented my description of neutrino entrons and neutrino photons in patent applications that I have filed. These applications have been available to everyone on earth via the United States Patent and Trademark Office web site for several years. If the reader is interested he can go to www.uspto.com and search for "tronnies".

The Proton's Mass

One of these neutrino entrons, captured by an electron, is located in each proton in our Universe and provides the proton with almost all of its mass. Similarly, one of these neutrino entrons, captured by a positron, is a component of each anti-proton and provides the anti-proton with almost all of its mass. Anti-protons are the anti-particle of the proton. Protons and anti-protons are described in **Chapter VIII**. Anti-protons are extremely rare in most parts of our Universe but they are produced in Black Holes at the center of each galaxy. (The Ross Model of Black Holes is described in **Chapter XX**.)

Neutrino Photons are the Carriers of Gravity

There is a Black Hole at the center of every galaxy. Protons and anti-protons are destroyed in Black Holes by their combination which annihilates both the proton and the anti-proton and releases the two neutrino entrons along with three electrons and three positrons. Many of these neutrino entrons escape from the Black Holes as neutrino photons and spread out through the galaxy surrounding the Black Hole in all directions. Nearly all of them that illuminate objects such as stars and planets pass through the objects applying a reverse Coulomb force on the objects directed back toward the Black Hole. So neutrino entrons (one in each neutrino photon) are the carriers of gravity as will be explained also in **Chapter XX**. They are the "gravitons" scientists have been for centuries searching for. Neutrino entrons in the neutrino photons are also the "Dark Matter" of our Universe providing the great majority of the mass in our Universe.

Neutrino Photon Speed is Always Constant

According to the Ross Model typical photons (such as photons in a visible light beam) slow down or speed up in order to travel at the speed of light through Coulomb fields that may be traveling at a variety of speeds relative to the source of the visible light beam and the universal Coulomb field. However, neutrino photons always travel at a constant speed equal to the speed of light relative to the speed of the source of the neutrino photons. This is important when we are analysing Black Holes and gravity. As will be explained in detail later on in **Chapters XX, XXIII and XXV**, celestial objects even galaxies can exceed the speed of light in the process of being pulled into Black Holes by the passage through them by neutrino photons. Visible light photons attempting to escape the Black Holes are forced to travel backwards toward the Black Hole in order to travel at the speed of light through the Coulomb fields of celestial objects incoming at speeds exceeding the speed of light. However, the tiny, extremely energetic neutrino photons continue to pass through the Coulomb fields of the incoming celestial objects at the speed of light relative to the Black Hole.

Pair Production and Electron-Positron Annihilation

A neutrino entron combines with a gamma ray entron and a low-energy entron in the formation of each electron-positron pair in a process called "pair production". (Pair production is a well-known process described briefly in **Chapter I**, but existing descriptions of the process do not include the neutrino entron or any other entron.) According to this version of the Ross Model the 928 MeV neutrino entron, as explained above, has a diameter of about 0.934×10^{-18} m and its photon circle has a diameter of about 0.85×10^{-15} m. The diameter of the entron circle of the 1.02 MeV gamma ray photon, also is about 0.85×10^{-15} m, and the 1.02 MeV photon diameter is about 7.7×10^{-13} m. The 1.12 KeV entron has a diameter of 7.7×10^{-13} m. Therefore at locations where there is a large flux of 1.12 KeV photons, a large flux of 1.02 MeV gamma ray photons and a large flux of 928 MeV neutrino photons a combination of one 928 MeV neutrino photon, one 1.02 MeV gamma ray entron and one 1.12 X KeV entron is possible. When that combination occurs, the neutrino entron with its enormous mass dominates the combination and the result is that its two tronnies continue in their 0.934×10^{-18} m circle and the four tronnies of the other two entrons circle through the center of the neutrino entron circle. Then each of the two tronnies of the neutrino entron (one plus and one minus) picks two of the four tronnies with opposite charge and the two sets of three tronnies scoot off in opposite directions as an electron and a positron.

In this process the neutrino entron and the other two entrons disappear. However, a new neutrino entron is produced and released in a subsequent process called "electron-positron annihilation" when the positron produced in the pair production process combines with an electron. Two 0.51 MeV entrons are also released in the annihilation process. The two 0.51 MeV entrons can be detected. To my knowledge the neutrino entron in the past has not been detected. In the past no one has known that it exists. If and to the extent it has been detected, no one has known what it was.

Forces within the Neutrino Entron

Neutrino entrons have a diameter estimated by me to be 0.934×10^{-18} m as explained above, so we can, using **Equation 2**, estimate the attractive and repulsive forces acting between the two tronnies of the neutrino entron. According to Coulomb's Law, if these two tronnies were stationary and separated by the diameter of the neutrino entron, the attractive force applied to each of the two tronnies on each other would according to **Equation 2** be:

$$F = kQ_1Q_2/r^2 \qquad (2)$$

where k is 8.99×10^9 Nm^2C^2, Q_1 and Q_2 are each equal to 1.602×10^{-19} C, and r is the diameter of the neutrino entron's circle, 0.934×10^{-18} m. Since r in this case is the diameter d' of the entron circle, let's replace the r with a d' to avoid confusion so:

$$F = kQ_1Q_2/(d')^2 \qquad (2A)$$
$$F = (8.99 \times 10^9 \ Nm^2C^2)(1.602 \times 10^{-19C})^2/0.934 \times 10^{-18} \ m$$

If you do the math you will see that the attractive force on each tronnie from its partner on the opposite side of the neutrino entron would be about 265 million newtons, again assuming that the two charges were stationary. This would be a tremendous force equivalent to about 29 thousand tons. However, the charges are not stationary, not by a long shot. They are circling at about 1.5708 times the speed of light as shown in **FIGS. 2A and 2B**. Tronnie P at position 200 feels a repulsive force from itself coming from position 202 (a distance d' = 0.934×10^{-18} m) and feels an attractive force from Tronnie N coming from position 204 (a distance of 0.55536×10^{-18} m, a distance shorter than a billion-billionth of a meter). So what we need to do is to integrate the forces around the respective circles (one larger and one smaller) defined by the distances traveled by the Coulomb forces between each tronnie and itself (200 to 202, this is the larger circle with diameter d') and between each tronnie and the other tronnie (204 to 200 and 206 to 202, these are the smaller circles with diameter 0.5946 d').

The repulsive Coulomb forces travel across the diameter of the respective circles while the tronnies travel around the circumference. When these forces are integrated for one cycle, the integrated forces are:

$$F_I = \int_0^C \frac{kQ_1Q_2}{(d')^2}$$

So we are integrating around the circle along the circumference from zero to C. Recognizing that $C = \pi d$ so d' can be replaced by C/π so:

$$F_I = \int_0^C \frac{kQ_1Q_2}{C^2/\pi^2} dC = \int_0^C \frac{\pi^2 kQ_1Q_2}{C^2} dC$$

π^2, k, Q_1 and Q_2 are all constant so these terms can be moved outside the integral, so:

$$F_I = \pi^2 kQ_1Q_2 \int_0^C \frac{dC}{C} = \pi^2 kQ_1Q_2 \int_0^C C^{-2} dC$$

From a table of integrals I see that:

$$\int x^n dx = \frac{x^{n+1}}{n+1} \text{ so} \int C^{-2} dC = \frac{C^{-2+1}}{-2+1} = \frac{C^{-1}}{-1} = C^{-1}$$

So:

$$F_I = \pi^2 kQ_1Q_2C^{-1} = \frac{\pi^2 kQ_1Q_2}{\pi d'}$$

since $C^{-1} = 1/\pi d'$, and we can cancel π in the denominator and one π in the numerator, so

$$F_I = \pi kQ_1Q_2/d' \qquad (3)$$

Equation 3 shows that the integrated forces are inversely proportional to d' instead of $(d')^2$ as in **Equation 2A**. Also, since k has units of Nm^2/C^2, Q_1 and Q_2 have units of coulombs and d has units in meters, the integrated forces have energy units of joules (or electron-volts) rather than force units of newtons as in **Equation 2**.

When we plug in the numbers we see that the repulsive integrated forces in the diametrical direction for the neutrino entron are equal to approximately 7.76×10^{-10} joules (4.84×10^9 eV). The attractive integrated forces are equal to approximately 13.0515×10^{-10} joules (8.18×10^9 eV); however, the attractive integrated forces in the diametrical direction are this force decreased by the cosine of 53.515 degrees which is 0.59461. So the attractive integrated forces in the diametrical direction are (0.59461) X (13.0515×10^{-10} joules) = 7.76×10^{-10} joules which is exactly equal to the integrated repulsive forces in the diametrical direction. The attractive integrated forces in the tangential directions are 13.0515×10^{-10} joules multiplied by the sine of 53.515 degrees

50

(i.e. 0.804) which product is about 10.5 X 10^{-10} joules. Since everything in this entron is in perfect dynamic balance, the neutrino entron is an extremely stable particle.

Once created neutrino entrons exists forever except, as described above, they disappear briefly then reappear in connection with pair production and electron-positron annihilation processes. According to the Ross Model neutrino entrons are so stable that all of the neutrino entrons in our Universe will survive the ultimate destruction of our Universe and participate in all of our successor universes. Recycling of universes will be discussed in detail in **Chapter XXV**.

(You the reader may be curious to know how these integrated forces of the neutrino photon, attractive and repulsive, which have the units of energy (joules), relate to the energy of the neutrino entron and the neutrino photon which I have estimated to be about 9.28 X 10^{8} eV which is about one fifth of my estimate of the attractive and repulsive diametrical integrated force of 48.4 X 10^{8} eV. I am similarly curious but so far I have not figured out the correlation. It may be that I have one or more errors in my calculations or assumptions. I suspect that the lack of perfect correlation results from the fact that the "energy of the neutrino entron" estimated at 1.487 X 10^{-10} J (9.28 X 10^{8} eV) is not really the energy of the neutrino entron but is the energy that the neutrino entron supplies to the particle that captures it, such as the electron in the proton. I challenge the reader to help me out. And if I am notified of the answer, I will include it in the next version of the Ross Model.)

The reader should recognize that the development of **Equation 3** above was done using the neutrino entron as an example with d' equal to 0.934 X 10^{-18} m; however, the equation applies for all entrons. To determine the integrated forces acting within an entron such as any one of the examples in **Table V**, merely substitute the appropriate diameter d'. For example the repulsive integrated force acting on each of the two tronnies in the 2.29 eV green light entron with a diameter d' of about 3.77 X 10^{-10} m is about 12 eV. The attractive force in the diametrical direction is also about 12 eV. This corresponds to the recognized energy of the green light photon, which is 2.29 eV. So our calculation of attractive and repulsive forces in the diametrical direction exceeds the recognized energy of the green light photon by a factor of about 5.2. There may be a way to correlate the energy of entrons with the integrated forces within the entrons, but I have not yet discovered what that correlation is. I feel lucky that the values are in the same ballpark.

Importance of the Neutrino Entron

Neutrino entrons and neutrino photons were unknown to science until I discovered them in the course of developing my model of our Universe. Our Universe would not exist without these two simple particles, each comprised of two opposite tronnies. Without them there would be no electrons, no positrons, no protons, no gravity, no atoms, no molecules, no stars and no galaxies. Without neutrino entrons and neutrino photons, our Universe would be comprised of only tronnies, Coulomb force waves and lower energy entrons in the form of photons. I suspect that prior to the creation of the first universe similar to our own, empty space was filled with only Coulomb force waves, tronnies and lower energy entrons in the form of photons.

Granddaughter, Kate Ross, with the electron-positron TinkerToy® model representing eight snapshot positions of the three plus and minus tronnies in the electron or the positron.

Electrons and positrons are each the combination of three tronnies.

In an electron a plus tronnie travels in a circle with a diameter of 0.934×10^{-18} m and two minus tronnies travel in the same size circle around the path of the plus tronnie, 90 degrees behind the plus tronnie. Take a look at FIG. 5.

Most electrons and positrons exist in their natural state that Mr. Ross calls their ground state, and he refers to these ground states as naked electrons and naked positrons. Naked electrons and naked positrons are self-propelled by internal Coulomb forces produced by their circling tronnies at a speed of 2.19×10^6 meters per second. Their natural speed combined with their masses gives the electrons and the positrons a kinetic energy of 13.6 eV, but these naked electrons and naked positrons have no electrical energy. Electrons can capture (or be captured by) entrons to become energetic electrons and these electrons will have electrical energy and can be manipulated to power and light our civilazations.

CHAPTER VII

ELECTRONS AND POSITRONS

Naked Electrons

To the best of my knowledge, the Ross Model describes for the first time the internal structure of electrons and explains for the first time how internal Coulomb forces within electrons provides a self-propulsion for the electron at a speed close to one per cent of the speed of light. These electrons in their ground state are referred to in the Ross Model as "naked electrons". **FIG. 5** is a snap-shot drawing of a naked electron. Naked electrons are comprised of one plus tronnie shown as a plus sign at 302, traveling in a circular path 300 with a diameter of 0.934×10^{-18} m at a speed of $\pi c/2$ (about 1.57 times the speed of light). And the two minus tronnies, each shown as a minus sign at 304A and 306A, are circling the path of the plus tronnie in circles of the same diameter, speed and frequency of the plus tronnie. At the instant of the **FIG. 5** snap shot drawing, the two circling minus tronnies are circling around location 302A in a path perpendicular to the path 300, one-fourth period behind the plus tronnie which is as shown in **FIG. 5** now located at location 302. The time for the three tronnies of the electron to complete one cycle is about 0.622×10^{-26} second. The electron's frequency is about 1.606×10^{26} cycles per second (that is about 160 trillion-trillion cycles per second). That is a lot of spinning! The reader should recognize here that the naked electron is extremely small with a shape defined by the paths of its three tronnies. The electron would fit inside a cubic box with edges of 2×10^{-18} m. Except for tronnies (which have no size at all) and neutrino entrons, with dimensions about one half the size of electrons, the naked electron and its anti-particle, the naked positron, are the smallest things in our Universe. Typical atoms are about 100 million times larger than naked electrons and positrons. Naked electrons and naked positrons are self-propelled. Note that in the drawing of the naked electron, both of the minus tronnies pass upward through the center of circle 300 at the rate of 1.6×10^{26} (160 trillion-trillion) passes per second. On each pass through the center at speeds faster than the speed of light, Coulomb force waves from the two minus tronnies attract the plus tronnie in an upward direction giving the naked electron a natural velocity. According to the Ross Model that natural velocity is currently estimated to be 2.19×10^6 m/s (2.19 million meters per second) a little less than one percent of the speed of light) giving the naked electron a kinetic energy ($E = mv^2/2$) of 2.18×10^{-18} joules, i.e.:

$$E = (\tfrac{1}{2})mv^2 = \tfrac{1}{2}(9.102 \times 10^{-31} \text{ kg})(2.19 \times 10^6 \text{ m/s})^2$$
$$E = 2.18 \times 10^{-18} \text{ kgm}^2/\text{s}^2 = 2.18 \times 10^{-18} \text{ joules.}$$

FIG. 5

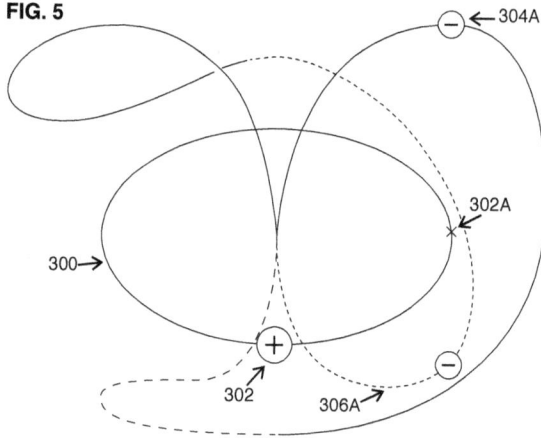

FIG. 6

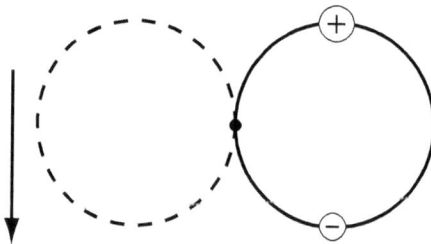

An energy of 2.18 X 10^{-18} joules is equivalent to 13.6 electron volts. According to the Ross Model, in atoms all of the orbital electrons in their ground state are naked electrons circling around atomic nuclei at their natural speed of about 2.19 X 10^6 m/s, a little less than one per cent of the speed of light. This explains why orbiting electrons do not lose energy as they orbit. They are naturally self-propelled by their own internal Coulomb forces giving them a kinetic energy of about 2.18 X 10^{-18} joules. This is energy that they cannot lose. They can remain in orbit around the positive nuclei of atoms for billions of years without being drawn into the nucleus by attractive Coulomb forces. (Some of the electrons may move at an average speed of about 2.19 X 10^6 m/s in paths other than circular as a result of outside Coulomb forces, such as periodic forces produced by other close-by electrons or protons.)

Some of you readers may remember from your high school chemistry or physics Niels

Bohr's description of the hydrogen atom in which his calculation showed that the electron's speed circling the proton nucleus was about 2.19×10^6 m/s. His model of the hydrogen atom has been largely discredited by modern physicists. Professor Bohr did not suggest that electrons were self-propelled. My model of the electron clearly shows that it is self-propelled.

The Reader should understand what I mean by the various forms of energy of electrons. Although, according to the Ross Model, naked electrons have a tremendous natural velocity, about 2.19×10^6 m/s (about 2.19 million meters per second), giving each electron a kinetic energy of 13.6 eV; the naked electron carries no electrical energy. Electrical energy of the electron is determined by the energy of any entrons captured by the electron. The electrical energy of naked electrons is zero, because it has captured no entrons! This is an extremely important revelation of the Ross Model. Almost all electrons in orbit around the nuclei of atoms are naked electrons with no electrical energy. On the other hand conduction electrons (in metallic and some other atomic structures) which move freely through the atomic structures (especially metallic structures) are almost always energetic electrons, and energetic electrons almost always carry electrical energy in the form of their captured entrons, as we will see in the next section. You will also discover in **Chapter XIV** that the Ross Model proposes that magnetic fields (including the earth's magnetic field) are comprised of nothing but naked electrons traveling through or through and around magnetic matter (such as iron, cobalt or nickel) at their natural speed of about 2.19×10^6 m/s. In an electric generator naked electrons in rotating magnetic fields can be used to force energetic conduction electrons in copper wire to oscillate back and forth in the copper wire as high-voltage alternating current.

Energetic Electrons

Naked electrons can capture entrons to become "energetic electrons". The Ross Model proposes two classes of energetic electrons, "low-energy energetic electrons" and "high-energy energetic electrons". Entrons with entron energies less than the electron's natural kinetic energy (less than about 13.6 eV or 2.18×10^{-18} J) slow the electron down. I refer to these electrons, (i.e. electrons having captured an entron with energies less than 13.6 eV) as "low-energy electrons". **FIG. 6** is a drawing of an energetic electron. Both the plus tronnie and the minus tronnie of the entron are passing upward through the electron applying a Coulomb force to the electron in a downward direction as shown in by the arrow in FIG. 6. When its electrical energy is less than 13.6 eV the kinetic energy of the electron is the difference between 13.6 eV and its electrical energy. An energetic electron with 13.6 eV of electrical energy has approximately zero kinetic energy but its captured entron has energy of 13.6 eV. The 13.6 eV entron has a diameter d' equal to about 0.64 X 10^{-10} m which is about one half the size of a typical atom and about 300 million times larger than the electron. Lower energy entrons have an even larger diameter. So the actual size of the electron is very greatly exaggerated in **FIG. 6** unless the energy of the entron is very large. For lower energy entrons the path of the tronnie through the electron will appear to the electron as an approximately straight line and will propel the electron in a direction opposite the naked electron's natural direction. Only for very high-energy entrons are the electrons driven in a circular path corresponding to the path of the tronnies in the entron.

The two tronnies (one plus and one minus) of the captured entron pass through the plus tronnie circle of the naked electron in the same direction as the minus tronnies of the naked electron. Once an entron is captured by a naked electron the entron becomes part of the electron and the electron becomes an energetic electron. (For very high-energy entrons such as the neutrino entron, it may be more appropriate to say that the electron becomes part of the entron.) The Coulomb forces of the captured entrons tend to push the energetic electron in a direction opposite the natural direction of the naked electron as shown in **FIG. 6**. Low-energy entrons slow down the naked electron from its natural velocity of about 2.19 X 10^6 m/s to a velocity of approximately zero for entron energies of 13.6 eV and high-energy entrons (entron with energies greater than 13.6 eV) drive the energetic electron in a direction opposite the normal direction of the naked electron. An electron with a captured entron with energies greater than 13.6 eV is referred to in the Ross Model as a "high-energy electron".

As explained above the frequency, energy and mass of the entron is inversely proportional to the entron's diameter. The captured entrons of both the low-energy electrons and the high-energy electrons give the electrons their electrical energy and adds mass to the electron. For example all of the electrons carrying current in the portion of a copper wire connected to the high voltage terminal of a 6-volt battery will have electrical energy of about 6 eV (1.08 X 10^{-18} J) and a velocity that is very slightly lower than the natural velocity of the naked electron. On the other hand all of the orbital electrons, there are 28 of them; (the 29th electron is the current carrying electron) in the copper wire are naked electrons in their "ground state", have captured no entrons, carry zero electrical energy and travel in synchronization with each other at their natural velocity of 2.19 X 10^6 m/s.

Electrons in a Copper Matrix

According to my Ross Model each copper atom in a copper metal matrix has a charge of plus 29 in its nucleus but only 28 electrons orbiting its nucleus. The orbiting electrons are all naked electrons orbiting at a speed of about 2.19 X 10^6 m/s. They are arranged with two electrons in a first orbital, eight electrons in a second orbital and eighteen electrons in a third orbital. Copper's 29th electron is free to move about randomly in the copper matrix as a conduction electron. Isolated copper atoms are ions having only 28 orbiting electrons. At speeds of 2.19×10^6 m/s the 29th electron is going too fast to remain attached to any individual copper atom so each individual copper atom is charged positive. Therefore, copper as it is found in nature, is almost always in its ionic form combined with a negative ion such a sulfur ion as copper sulfide. Copper can be refined from copper sulfide ore by an electrolysis process in which the copper sulfide is mixed with sulfuric acid. The copper is deposited out of the solution onto electrodes charged to about 2 to 4 volts. After copper atoms have been formed into a copper block, the 29th electrons are welcome to remain in the block to balance the charge of the metal block to zero. The 29th electrons however are not associated with any particular copper atom but instead travel freely in the copper matrix as a conduction electron. The electrical energy of the conduction electrons is determined by the energy of the entrons captured by the conduction electrons. Once a matrix of copper is created, the energetic electron will capture entrons in the matrix such as thermal entrons, so that the energy of the conduction electrons will correspond to the temperature of the copper metal. At room temperature of about 300 K (27 °C and 80.6 °F) the electrical energy of the 29th electron will be about 0.13 eV. (See **Table V** in **Chapter V.**) If the copper metal is made a part of a direct current electric circuit, conduction electrons (which have captured entrons with energies corresponding to the electric potential of the electric power source, such as a 6-volt battery) will flow through the copper metal at speeds close to one percent of the speed of light until they lose their energies. These electrons lose their energies very quickly after disconnection from the battery.

The entron energy of the energetic electrons may be converted into heat and photons as in the filament of an incandescent lamp. In a simple circuit comprised of a 6-volt battery, a 6-volt incandescent lamp and connecting copper wire, 6 eV energetic electrons flow through the copper wire from the high-voltage side of the battery to the filament of the lamp where the 6 eV entrons are lost from the energetic electrons as thermal entrons, some of which escape from the filament as visible light photons. On the low-voltage side of the filament all conduction electrons are at the potential of the low-voltage terminal of the battery. Very quickly an equilibrium condition will prevail where 6 eV electrons leave the high-voltage terminal of the battery pass through the copper wire and into the lamp filament where the 6 eV entrons are converted to thermal entrons increasing the temperature of the filament to a temperature approximately equal to the temperature of the surface of the sun. The electrons leave the filament as very low energetic electrons (about 0.13 eV) and continue through the copper wire to the low voltage terminal of the battery. The negative Coulomb forces provided by 6 eV electrons in the copper matrix prevent additional 6 eV electrons from leaving the battery and entering the copper wire on the high-voltage side of the circuit until an equal number of 0.13 eV conduction electrons exit the copper wire as low-voltage electrons at the low-voltage side of the circuit and enter the battery. No electrons are lost in the process. The entrons that are lost in the filament in the form of thermal entrons and photons are entrons that had been stored in the battery in the form of chemical energy.

In an alternating current circuit, electrons with energies ranging from zero to peak energies flow back and forth in the circuit in accordance with the frequency of the circuit. Once the power source is disconnected from the copper metal, any electrons then carrying high-voltage entrons will quickly share the entrons with other conduction electrons in the copper matrix and the entrons will be lost from the conduction electrons and will be converted to heat entrons increasing slightly the temperature of the copper metal for a while and ultimately released as thermal photons.

Zinc is similar to copper. This atom also has 28 orbital electrons and two electrons that are unwelcome in orbits around the zinc nucleus so they freely roam through the zinc metal matrix. Zinc is also normally found in nature as an ore such as zinc sulfide or zinc oxide and is refined similarly to copper in a process where low-voltage reduced-speed energetic electrons are added to naked zinc atoms. These reduced-speed electrons can remain in the metallic zinc matrix as conduction electrons. In zinc alkaline disposable batteries zinc's 29th and 30th electrons are each carrying energetic entrons giving each battery a potential of about 1.5 eV. Four of these batteries in series will provide a potential of about 6 volts. These metallic zinc atoms are converted to zinc oxide in these disposable batteries when they are oxidized in the battery leaving behind their two energetic electrons to provide electric energy in an electric circuit provided by the disposable batteries. I will have more to say about batteries and zinc in **Chapter XIV**.

Positrons

Positrons, naked and energetic (as in prior art models) are the anti-particle of electrons and naked positrons are also self-propelled at the same speed of 2.19 X 10^6 m/s. A drawing of a naked positron would be exactly like the **FIG. 5** drawing except the center tronnie is a minus tronnie and the two tronnies circling the center tronnie are plus tronnies. As I have explained most people are familiar with electrons and most people are not familiar with positrons. I also explained in **Chapter I** that one electron and one positron can be created in a process called pair production and the positron created will quickly combine with an electron, and the positron and the electron will annihilate each other in a process called electron-positron annihilation. This may cause you to wonder if there may be the same number of positrons in our Universe as there are electrons. If that is the case, then where are all the positrons hiding? I answer that question in the next chapter.

Tinker Toy® Model of Electrons, Positron and Proton

The reader should note that the positron has the same basic shape as the electron shown in **FIG. 5**. As you will see in the next chapter the proton also has the same shape as that of the electron but is much larger and much more massive. I have made a single model that represents each of these particles using Tinker Toy® parts. (See the photograph on page 53 of the model and my beautiful granddaughter Kate Ross.) The model is also a model of the combination of the tronnies of the three entrons that combine to produce an electron and a positron in pair production. And it is also a model of the electron-positron combination just prior to annihilation of the electron and the positron. Here is the process for making such a model for the ground state electron. First you will need to borrow some Tinker Toy® parts from the kid next door and purchase some 3/8-inch flexible plastic tube from a hardware store.

Parts needed:
1) Nine yellow 2-inch diameter 1-inch wide socket parts each having eight 3/8 inch slots in its circumference and two 3/8 inch slots located at the center of both of its sides.
2) Twenty four 4 and ¾ inches long 3/8 inch diameter red rods
3) Fourteen 1 ½ by 2 inch purple cross pieces each having 3/8 inch diameter slots at the two ends of one of the cross pieces and a 3/8 inch hole through the other cross piece.
4) Approximately 12 feet of flexible plastic 3/8 inch diameter plastic tubing, cut into eight 4.25-inch pieces and two 5-foot sections. Trim a ½ inch section at both ends of the 5-foot sections so that the ends of both 3/8 inch tubes will fit into the two 3/8 inch slots located at the center of both of sides one of the yellow socket parts.

Instructions:
1) Use one of the yellow socket parts to represent the center of the plus tronnie circle.
2) Insert a red rod in each of its eight circumference slots.
3) Attach the other eight yellow socket parts to the other ends of the red rods.
4) Position the eight yellow socket parts orthogonal to the plane of the red rods.
5) Insert a first red rod in one of the yellow socket part in a direction parallel to the red rod that is inserted in the center socket.
6) Moving counter clockwise around the eight yellow socket parts, insert in the adjacent socket part two red rods parallel to each other in directions 45 degrees relative to the plane defined by the eight red rods inserted in step 2.

7) Moving counter clockwise repeat step 6 advancing the directions of the red rods by 45 degrees (relative to the plane of the eight inserted red rods) with each step until the eight socket parts have two red rods positioned (beginning with the first red rod and going counter clockwise) at 0, 45, 90, 45, 0, 45, 90, 45 and back to 0 degrees, these rods defining fourteen extended red rods.

8) Using one of their 3/8 inch diameter slots, attach a purple cross piece at the unattached end of each of the fourteen extended red rods.

9) Connect each of the eight other yellow socket parts with the eight 4.25-inch pieces of plastic tube using the side 3/8 inch slots of the yellow socket parts.

10) Weave the two five-foot sections of flexible plastic tubing through the holes in the purple cross pieces and insert the four ends of two tubes into the two slots of the center socket part.

With respect to the electron, the Tinker Toy model represents eight snapshots of the electron during one cycle of the electron. The eight other yellow socket parts represent eight positions of the plus tronnie during one of the electron's cycle. The eight 4.25-inch pieces of plastic tubing represents the path of the plus tronnie during one electron cycle. Each of the two five-foot sections of tubing represents the path of one of the two minus tronnies during one electron cycle. The fourteen purple cross pieces plus the center yellow socket part represent sixteen positions of the two minus tronnies during one electron cycle. Scientists are aware that the electron has a property called "spin"; however, some believe that its spin is not real. The Ross Model clearly shows that the spin is real. It is amazingly real. The time for each cycle is 0.6227×10^{-26} second. The electron's frequency is about 1.6059×10^{26} cy/s **(about 160 trillion-trillion cycles per second!)**. This extreme frequency results from the fact that the diameter of the plus tronnies circle is only about 0.933×10^{-18} m and the Coulomb forces driving the tronnies are traveling at the speed of light, about 3×10^8 m/s. Tinker Toys are perfect for modeling an electron since the eight slots in the circumference of the yellow socket part are raidally separated by 45 degrees making it easy to position the two minus tronnies 90 degrees behind the plus tronnie, perpendicular to the path of the plus tronnie and circling with the same diameter as the diameter of the plus tronnie. See the photograph on page 53.

This model also represents the path of the minus tronnie and the two plus tronnies that comprise each positron. The model also represents the path of the energetic electron and the two positrons that comprise the ground state proton that I will describe in the next chapter.

In addition the model also represents the tronnies of the three entrons (a neutrino entron, a 1.02 MeV gamma ray entron and a low-energy entron) that are joined together in pair production events to produce an electron-positron pair. The fourteen purple pieces and the center yellow socket part represent four positions of the two tronnies of the 1.02 MeV gamma ray entron and the two tronnies of the low energy entron. The two tronnies of the neutrino entron are on the opposite sides of the yellow socket part circle, and just before the electron and positron split off, two minus tronnies are circling 90 degrees behind the plus tronnie of the neutrino entron and two plus tronnies are circling 90 degrees behind the minus tronnie of the neutrino entron. Then the electron and the positron break apart and the electron and the positron (with a total of six tronnies) are created from the six tronnies of the three entrons. A similar but reverse process takes place in electron-positron annihilation processes, in which two 0.51 MeV entrons and a new neutrino photon is produced with the demise of the electron and positron.

Models of Perpetual Motion Machines

As explained above I have for several years been attempting without success to patent processes for making models of things described in this book. The United States patent office generally does not permit inventors to submit models of their invention; but the USPTO rules (see MPEP 608.03) do permit models of perpetual motion machines to be admitted as a part of an application for patent, but the patent office has never granted a patent on a perpetual motion machine. I have encouraged the Examiner to construct the model described above which I think is a perpetual motion machine and I have offered to mail an already constructed model to the Examiner. I am not attempting to patent these particles, but I am attempting to patent processes for making models of them.

Our Cold Plasma Shell

As explained above naked electrons and naked positrons travel at a speed of 2.19×10^6 m/s. Having opposite charges of plus and minus e, these particles are very attractive to each other, especially at close ranges, but they also repel themselves with their own internal Coulomb forces. These particles are also extremely small at about 2×10^{-18} m. As a result huge numbers of these particles can exist as a cold plasma. As you will see in **Chapter XXII**, the Ross Model proposes that naked electrons and naked positrons form a cold plasma shell many light years thick surrounding our Universe. In this shell the naked electrons and naked positrons travel randomly at 2.19×10^6 m/s. At this speed and with their extremely small size, collisions between the particles are extremely rare. Since both electrons and positrons are charged particles they reflect low energy electromagnetic radiation creating the uniform cosmic background radiation. See **Chapter XXII**.

Ground state protons are created by the combination of an electron, a neutrino entron and two positrons. Every proton created reduces the free positron population in our Universe by one as compared to the number of electrons. The neutrino entron increases the electron mass from about 9.1×10^{-31} kg to about $16,549.7 \times 10^{-31}$ kg.

FIG. 7 is a drawing of a ground state proton (also called a "naked proton"). It has a natural velocity of 4.02×10^7 m/s and a kinetic energy of 8.37 MeV but no gamma ray energy. In the early phase of our Universe the naked protons captured gamma ray entrons to slow down to speeds close to zero, cooling down our Universe in the process. The gamma ray entrons increase slightly (about one hundredth of one percent) the proton's mass. The total energy of its gamma ray entrons is about 8.37 MeV equal to about 1.674×10^{-31} kilograms.

When four protons and two electrons combine to form an alpha particle in a fusion process, some of the gamma ray entrons are released as fusion energy. This is how our sun's energy is produced explaining why hydrogen bombs are so destructive. We will learn in Chapter XX how proton destruction in Black Holes produces the gravity holding galaxies together and you to the chair you are sitting on.

CHAPTER VIII

PROTONS

Naked Protons and Energetic Protons

According to the Ross Model there are, like electrons, two types of protons, "naked protons" and "energetic protons". An energetic proton is a naked proton that has captured at least one entron.

The Creation of Naked Protons

According to the Ross Model each naked proton contains two naked positrons (each with a tiny mass of 9.109×10^{-31} kg), and one very energetic electron that is a naked electron that has captured a neutrino entron (with a mass/energy of about $16,549.7 \times 10^{-31}$ kg, more than a thousand times greater than the mass of the electron) to form a very massive $16,558.8 \times 10^{-31}$ kg (about 1.65588×10^{-27} kg) electron traveling in a circle at a speed of about 1.57 times the speed of light. **FIG. 7** is a snap-shot graphic model of a naked proton. The massive electron accounts for about 99.9 percent of the naked proton's mass. The two positrons account for only about 0.1 percent of the naked proton's mass. The net charge of the naked proton is +1e as a result of the combination of the charges of its two positrons (+2e) and its one high-energy electron (-1e). The neutrino entron (which is part of the electron) has one plus charge and one minus charge and a net charge of zero. (We can also count the tronnies in the naked proton, three in each of the two positrons, three in the naked electron and two in the neutrino entron, so the naked proton is comprised of six plus tronnies and five minus tronnies which also result in a net charge of +1e.)

According to the Ross Model the frequency of the neutrino entron is exactly the same as that for the electron, and the diameter of the neutrino entron is 0.934×10^{-18} m equal to the diameter of the path of each of the tronnies in the naked electron, about one-half the size of the naked electron which has a shape shown in **FIG. 5**. The electron has a natural speed of about 2.19×10^6 m/s and the neutrino entron as the only component of the neutrino photon, changes speed (relative to the space through which the photon is traveling) during each photon cycle from about minus 3×10^8 m/s to about plus 9×10^8 m/s as indicated in **FIG. 4**. We will see in **Chapter XX** that I have estimated that the neutrino entron flux passing through earth from the Black Hole in the center of the Milky Way Galaxy could be about 300,000 photons/m^2-second. So in the very unlikely event that the neutrino entron (as the only component of the neutrino photon) resonantly intersects with an electron with each of them traveling at similar speeds; there is a good chance that the neutrino entron will be captured by the electron. (Or we could say that the neutrino entron captures the electron.) When that happens, the entron becomes a part of the electron and increases the electron's velocity to a speed greater than the speed of light (about 1.414 c as explained in more detail below). However, instead of driving the electron in a straight line as might be expected, the entron drives the electron in the neutrino photon's circle (see **FIG. 3**) which, as explained above, is a circle with a diameter of 0.6366λ where λ is the wavelength of the neutrino photon. The wavelength of the neutrino photon is about 1.34×10^{-15} meters, so the diameter of the circle of the electron and the path of its captured neutrino entron are both about 0.85×10^{-15} meters. If there are positrons in the vicinity of the rapidly circling electron, they will be attracted to the center of the negative energetic electron's circle and begin circling behind the energetic electron as shown in **FIG. 7**. The very energetic negative electron circling in a tiny circle with a diameter of 0.85×10^{-15} m at 1.414c will exert a Coulomb force on itself across the diameter of the circle and this additional repulsive force will boost the speed of the very energetic electron to $\pi c/2$ (about 1.57c) which adds stability to the proton. If one positron attempts to combine with the spinning electron the combination will be unstable; however, two positrons in combination with the spinning neutrino-entron-energized electron form the most stable composite particle in our Universe (i.e. the naked proton). The two naked positrons each with a mass of only 9.109×10^{-31} kg are circling the path of the very energetic electron at a location one-fourth period behind the electron. The diameter of the circle of each of the naked positrons is also 0.85×10^{-15} m to give the naked proton a size of about 1.7×10^{-15} m. The period of the naked proton (i.e. the time for one orbit of the electron and the two positrons is 2d/c which is about $(2)(0.85 \times 10^{-15}$m$)/(3 \times 10^{-8}$ m/s$)$ = about 0.5667×10^{-23} second, which is equivalent to a frequency of about 1.765×10^{23} cycles per second (about 0.1765 trillion trillion cycles per second. And some people think the proton spin is a fiction!

As will be explained in detail later on, this creation of protons from the combination of an electron, two positrons and a neutrino entron is not currently happening on any significant scale, in the part of our Universe that is visible to us. This is because there are not many free positrons available in this portion of our Universe. Almost all of them are already contained in protons. (The reader should keep in mind that we cannot see what is going on inside of Black Holes at the center of every galaxy or in the Shell of our Universe. Black Holes as discussed in **Chapter XX** and the Shell of our Universe are discussed in **Chapter XXII**.) However in the very early seconds after the Big Bang there were almost exactly as many positrons as there were electrons and a tremendous flux of neutrino photons.

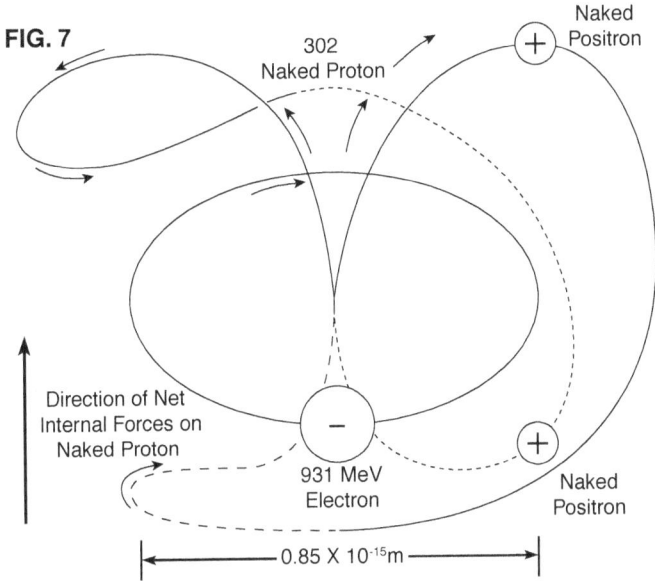

FIG. 7

302 Naked Proton

Naked Positron

Direction of Net Internal Forces on Naked Proton

931 MeV Electron

Naked Positron

0.85 X 10⁻¹⁵m

So soon after the Big Bang huge numbers of neutrino entrons and electrons combined to produce huge numbers of massive electrons. And the combination of each massive electron and two naked positrons produced huge numbers of protons during the early phase of our Universe. As we will see later on more protons are currently being created within the plasma shell of our Universe which is comprised mostly of high-speed electrons and positrons. Also, according to the Ross Model, huge numbers of anti-protons are currently being created inside Black Holes, but each of these anti-protons quickly combines with a proton and both the proton and the anti-proton are annihilated releasing their two neutrino entrons. But we will get to Black Holes, the formation of our Universe and the description of its shell later in this book. **FIG. 7** is a drawing showing the structure of the naked proton. Its structure is similar to the structure of the naked electron. The spin of the proton is similar to the spin of the electron, but the proton is about 1800 times more massive than the electron and about 580 times larger than the electron.

Naked Proton's Natural Velocity

According to the Ross Model the naked proton has a natural velocity, like the naked electron that is a fraction of the speed of light but the naked proton's natural speed is much faster than the speed of the naked electron. **That speed is so fast that the naked proton cannot capture an electron to become a hydrogen atom!** The naked proton must first capture gamma ray entrons to slow down. When a naked proton captures one or more gamma ray entrons it becomes an energetic proton. To estimate the energy/mass of the gamma ray entrons needed by the naked proton to slow down to approximately zero, the Ross Model compares the iron-56 atom to the hydrogen-1 atom which is comprised of only one proton, one orbiting electron and, according to the Ross Model, several gamma ray entrons circling through the proton. The Ross Model assumes that the iron-56 atom is comprised of no significant mass of gamma ray entrons. The Ross Model assumes that the proton of the hydrogen-1 atom has captured entrons having sufficient total energy to slow the proton down to a speed close to zero. The question is: "What are the mass and energy of these entrons?"

The Ross Model assumes that the nuclei of all stable neutral atoms (other than hydrogen-1 and possibly iron-56) are comprised of naked protons, naked electrons and high-energy entrons in their nuclei. These stable atoms also have a number of orbiting electrons (or orbiting electrons and conduction electrons) all or nearly all of which are naked electrons. (Hydrogen-1 has no electrons in its nucleus and iron 56 may or may not have any entrons in its nucleus. In the hydrogen-1 atom and in the iron-56 nucleus, we are not counting the electron, two positrons and the neutrino entron that are components of each naked proton.) The number of electrons in the nucleus of each atom is equal to the difference between the number of protons in the nucleus and the number of electrons associated with the nucleus. (According to the Ross Model there are no neutrons in the nuclei of atoms or if there are any, they each decay into a proton and an electron within a few minutes as they do outside atomic nuclei.) The hydrogen-1 atom (with one naked proton and an unknown number of captured entrons in the nucleus and one naked electron in orbit) has a mass of about 1.007828 atomic mass units (amu). An amu is equal to 1.66054×10^{-27} kg. (These numbers are from Table III.) So the mass of the hydrogen-1 atoms in kilogram units is about 1.673538×10^{-27} kg. The iron-56 isotope has the smallest mass-to-atomic number ratio of any stable isotope. The mass of the iron-56 isotope is about 55.934942 amu. Its atomic number is 56. The Ross Model assumes that the iron-56 atom is comprised of 56 naked protons and 56 naked electrons (30 of which are in the nucleus and 26 are outside the nucleus) and no entrons (or only entrons with insignificant mass/energy) within or outside the nucleus. With these assumptions the combined mass of one naked proton and one naked electron must be no greater than: 0.99883825 amu, i.e.:

$$55.934942 \text{ amu}/56 = 0.99883825 \text{ amu.}$$

Since 1 amu is equal to $1.6605402 \times 10^{-27}$ kg, this mass (of the naked proton and one naked electron) in kilograms is 1.658611×10^{-27} kg.

The mass of the hydrogen-1 atom is significantly greater. **The difference in mass between the hydrogen-1 atom and the combination of one naked proton and one naked electron must be equal to the mass of the entrons in the hydrogen-1 nucleus.** This leads in the Ross Model estimate that the entrons in the nucleus of the hydrogen-1 atom must have a mass equal to or greater than 0.00899 amu (i.e. 1.007828 amu minus 0.998838 amu). This difference of 0.00898 amu is equal to 0.0149×10^{-27} kg. (The mass of the entrons could be greater than about 0.0149×10^{-27} kg if the iron-56 nucleus contains one or more entrons with a total entron mass that is significant compared to 0.0149×10^{-27} kg.) The energy equivalent to this entron mass of 0.0149×10^{-27} kg is about 1.34×10^{-12} J (equivalent to 8.37 MeV). This is based on the well-known relation: $E = mc^2$. So:

$$E = (0.0149 \times 10^{-27} \text{ kg})(3 \times 10^8 \text{ m/s})^2$$
$$= 1.341 \times 10^{-7} \text{ J} = 8.37 \text{ MeV}$$

A single entron with this energy would have a wavelength of:

$$\lambda = hc/E = (6.626 \times 10^{-34} \text{ Js})(3 \times 10^8 \text{ m/s})/ 1.341 \times 10^{-12} \text{ J}) = 1.482 \times 10^{-13} \text{ m}$$

According to the Ross Model as explained in **Chapter XI**, the estimated diameter d' of an entron is related by the following formula to the wavelength λ of a photon that would be formed by the entron:

d' = $\lambda/1431$.

So the diameter of a single entron with an energy of 8.37 MeV would be:

d' = 1.482×10^{-13} m/1431 = 1.036×10^{-16} m.

This diameter appears too small relative to the diameter of the circle of the very energetic electron in the naked proton which is about 0.84954×10^{-15} m. Therefore, the Ross Model proposes that the naked proton is slowed down to near zero speed with the capture of several entrons (probability about 10 to 20) with energies in the gamma ray range. The total energy of all of these entrons is estimated to be about 1.341×10^{-12} J (equal to about 8.37 MeV). For example, if the number of entrons is 10, each with about the same energy, the diameters of these ten entrons would be about 1×10^{-15} m (a little larger than the diameter of the circle of the 931 MeV electron in the proton (see **FIG. 7**). If there are 20 approximately equal entrons slowing down the naked proton, their diameters would be about 2 X 10^{-15} m. If there are 15 gamma ray entron circling thorough the proton to slow it down to approximately zero speed. Each entron would have a diameter of about 1.5 X 10^{-15} m, an energy of about 0.558 MeV and a frequency of about 0.966 X 10^{23}/s which is about one-half the frequency of the proton. These diameters are larger than the naked proton but much smaller than the diameters of the orbits of the orbiting electrons which are about 1 X 10^{-10} m. (The Ross Model proposes that these orbiting entrons form a Coulomb barrier helping to prevent the orbiting negative electron of the hydrogen atom from being captured by the positively charged proton nucleus of the atom.) **Also very importantly some or all of these entrons are released as gamma ray photons in processes in which four protons and two electrons are joined together in a fusion processes to form an alpha particle. These released entrons represent the heat/energy of the hydrogen bomb and the heat/energy of our sun and most of the stars.** I will discuss these issues in more detail later on.

So now let us get back to estimating the natural, self-propulsion speed of the naked proton. This self-propulsion of the naked proton is due to the net Coulomb force applied by the two positrons in the naked proton as each of them pass upward through the center of the circular path of the proton's massive electron. According to the Ross Model that force applied twice during each of the proton's 1.786 X 10^{23} cycles per second, is sufficient to provide the naked proton with a natural velocity which is estimated using the equation: E = ½ mv^2, So,

$$v = \sqrt{\frac{2E}{m_n}}$$

where E is the total energy of the entrons needed to cancel the naked proton's natural kinetic energy and m_n is the mass of the naked proton. As explained above the Ross Model estimates the energy of the entrons needed to slow down the naked proton to about zero speed from its natural speed (that provides it with a kinetic energy of 8.37 MeV) is 8.37 MeV which is equivalent to 1.341 X 10^{-12} J. The estimated mass of the naked proton is 1.6586 X 10^{-27} kg, so the Ross Model's estimate of the natural velocity of the naked proton is:

$$v = \sqrt{\frac{2E}{m_n}} = \sqrt{\frac{2(1.341\ X\ 10^{-12}J)}{1.6596\ X\ 10^{-27}kg}} = 4.02\ X\ 10^7\ m/s$$

which is a little faster than **13 per cent of the speed of light!**

70

High-Velocity Protons –Hydrogen-1 Nuclei

The velocity of a proton, with 8.37 MeV of captured entron energy (also equal to 1.341×10^{-12} J) provided by a combination of gamma ray entrons (such as 15 gamma ray entrons) reduces the proton velocity to approximately zero as described above. Additional entrons can be provided to propel the proton in directions opposite the naked proton's natural direction as explained above with respect to the electron. However, the higher entron energies mean larger entron masses which increase the mass of the proton. When the entron mass/energy becomes large compared to the mass of the naked proton the proton's velocity levels off at 1.414 c as in the case of the energetic electron. The result is that the maximum velocity of a proton propelled with entron energy is the same as the maximum velocity of electrons propelled with entron energies, i.e. 1.414 c. **FIG. 8** in **Chapter IX** is a graph of electron and proton velocities as a function of captured entron energies. The formula for proton speed is the same as **Equation (9)** developed for the electron speed ion the next chapter (**Chapter IX**) except E_N for the proton is the natural energy of the naked proton and m_N is the mass of the naked proton.

The Anti-Proton

The anti-proton is exactly like the proton except in the anti-proton the positron has captured a neutrino entron and its path is circled by two naked electrons.

Destruction of Protons in Black Holes

Anti-protons are created in Black Holes when neutrino entrons are captured by positrons that are produced in the Black Holes with the destruction of protons and in electron-positron pair production processes. Then these anti-protons combine with protons to destroy in each case both the proton and the anti-proton, each releasing its neutrino photon and together releasing three positrons and three electrons to help maintain the cycle of proton destruction and neutrino entron release. As we will see in **Chapter XX** these neutrino entrons ultimately exit the Black Holes to produce the gravity holding the surrounding galaxy together.

All small particles are self-propelled. Entrons are self-propelled in the form of photons at the speed of light. Electrons are self-propelled at a speed of 2.19×10^6 m/s. Each naked proton is self-propelled at a speed of 4.02×10^7 m/s and must capture gamma ray entrons to slow down enough to become the nucleus of a hydrogen atom. Alpha particles also must capture entrons to slow down to become the nucleus of helium atoms. FIG. 8 shows the speed of electrons and protons as a function of the energy of their captured entrons.

Electrons and protons can be slowed down or speeded up by the capture of entrons. However, captured entrons add to the mass of the electrons and protons providing a speed limit for these particles. Particles approaching the speed of light have increased mass, not because they are going very fast but because mass has been added to the particles (in the form of entrons) to make them go very fast.

CHAPTER IX

SPEED OF SMALL PARTICLES

Self-Propulsion

Photons are self-propelled. Photons are each comprised of one entron traveling in a circle at a speed of 2c and forward at a speed of c (the speed of light) as explained in **Chapter IV**. Naked electrons and positrons are self-propelled by their own internal Coulomb forces at a speed of about 2.19×10^6 m/s, but they can capture low-energy entrons to slow down. Protons and alpha particles and atomic nuclei of small atoms are also self-propelled and slow down with the capture of high-energy entrons. Electrons and positrons can be propelled at high energies to speeds in excess of the speed of light by captured high-energy gamma ray entrons. Protons and alpha particles can also be propelled to very high speeds by captured gamma ray entrons. These high-energy entrons each has a mass that corresponds to their energy based on Albert Einstein's famous equation:

$$E = mc^2 \qquad (4)$$

Particles that capture entrons are slowed down or propelled by the captured entrons. The mass of the captured entrons add to the mass of the particles they are propelling. The speed of high-speed particles can only be increased with the capture of additional entrons which result in an increase in the mass of the particle. Entron masses are significant compared to the masses of subatomic particles only for very high-energy entrons. (See **Table V** in **Chapter V**.) Thus, the Ross Model, like Albert Einstein's special theory of relativity, provides an explanation as to why the mass of particles traveling close to the speed of light have substantially greater mass as compared to similar slow speed particles. This increase in mass with energy results in a limit on the speed of the energetic particles when the mass of the propelling entron becomes very large compared to the mass of the particle being propelled. In **FIG. 8** for example I have plotted the speeds of the electron and the proton as a function of the energy (in joules) of its captured entron. When the entron energy is low the speed of the particle is not significantly affected. When the entron energy is close to the natural kinetic energy of the particle, its speed approaches zero. Higher energy entrons result in increasing speed in the opposite directions. The net speed is determined by the well-known relationship:

$$E = (1/2)mv^2.$$

So:

$$v = \sqrt{\frac{2E}{m_n}}$$

But now the mass includes the mass of both the particle M_N and its captured entron m_e and the energy E is the difference between the natural energy E_N and the energy of the entron E_e. Therefore, the velocity of the particle is determined by the following formula:

$$v = \sqrt{\frac{2(E_N - E_e)}{M_N + m_e}} \qquad (9)$$

When the energy of the entron is small relative to the natural energy of the particle **Equation 5A** reduces to:

$$v = \sqrt{\frac{2(E_N)}{M_N + m_e}}$$

When the mass of the entron is small relative to the mass of the particle **Equation 5A** reduces to:

$$v = \sqrt{\frac{2(E_N - E_e)}{m_N}}$$

For the electron and the positron, the energy of the entron can be much greater than the natural energy of the particle and the mass of the entron can be much greater than the mass of the electron or the positron, so E_n and m_N can become negligible and the above equation reduces to:

$$v = \sqrt{\frac{2E_e}{m_e}} = \sqrt{\frac{2m_e c^2}{m_e}} = 1.414c$$

since the energy of the entron E_e is equivalent to $m_e c^2$.

Each entron captured by an electron adds to the mass of the electron, however for low-energy electrons the additional mass is insignificant. For example a 13.6 eV ultraviolet entron has a mass of only:

$$m_e = (13.6 \text{ eV})(1.784 \times 10^{-36} \text{ kg/eV}) = 2.43 \times 10^{-35} \text{ kg.}$$

This mass is insignificant compared to the zero energy electron mass of 9.109×10^{-31} kg. However, high-energy entrons can greatly increase the mass of electrons. For example a one million eV entron, which would be an entron produced by a one million volt accelerating source, would have a mass of:

$$m_e = (1 \times 10^6 \text{ eV})(1.784 \times 10^{-36} \text{kg/eV}) = 1.784 \times 10^{-30} \text{ kg.}$$

This additional mass is almost double the mass of the naked electron.

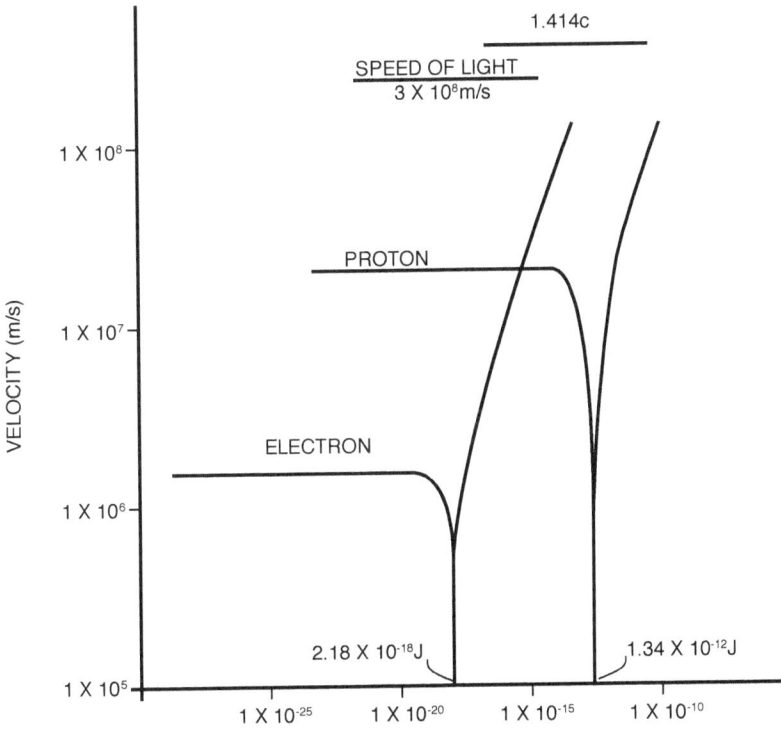

FIG. 8   ENTRON ENERGY (JOULES)

The plus and minus tronnies of each entron, as they pass through the capturing particle, apply a force on the particle in the direction opposite the particle's natural direction. As a consequence, at entron energies in the range of less than 13.6 eV (2.18×10^{-8}J) slow the electron down and the entrons with energies less than 6.24 MeV (6.66×10^{-13}J) slow the proton down. For higher energies, velocities increase fairly linearly with increasing entron energies but above a few MeV the velocity tends to level off as shown in **FIG. 8** as the mass of the entron approaches and exceeds the mass of the particle.

FIG. 8 shows electron and proton velocities as a function of entron energy. Also as the electron speed approaches the speed of light the diameter of the driving entron gets increasingly smaller causing the electron to move in circles that become smaller with increasing energy of the entron. Therefore, the two curves in **FIG. 8** are not continued in the region of the speed of light because the Ross Model proposes that very high-energy entrons drive the particles that have captured them in a circle the diameter of which is the same as the diameter a photon would have if the entron in question were traveling in the form of a photon at the speed of light. (As explained in **Chapter VIII** the high-energy electron in the Ross Model proton has a velocity of 1.57 times the speed of light, but its path is circular and the extra velocity (above 1.414c) is the result of the electron's internal Coulomb forces from across the diameter of the electron's circular path. Professor Einstein taught that a particle's mass increases because it goes fast making further acceleration more difficult, resulting in a speed limit of c. The Ross Model teaches that in order to drive a particle to high speeds in a single direction, we must drive it with entrons, but the driving entron adds mass to the particle and that increase in the mass of the particle makes it harder to increase the speed, resulting is a speed limit of 1.414c. But the direction of the entron driven particle becomes circular when driven by a single high-energy entron. Therefore, in order to drive an electron in a single direction faster than the speed of light, several lower energy entrons may be utilized. In any case there appears to be an upper limit on electron speed in a single direction of 1.414c.

Entrons Determine Speed of Energetic Electrons and Protons

The zero energy (naked) electron mass is assumed to be approximately the same as published values of mass of the electron at rest, i.e. 9.1094×10^{-31} kg as explained above (see **Table III** in **Chapter I**). The zero energy (naked) proton mass is determined by assuming that the iron 56 isotope is comprised of 56 protons (with no captured entrons having significant mass) and 56 electrons (with no captured entrons having significant mass) all as explained in **Chapter VIII**. (The reader should understand that when I am referring to zero energy naked particles, I mean that it has zero electrical energy (i.e. no captured entrons. In this context the zero-energy particles (naked particles) have very significant kinetic energy but no entron energy.)

We will learn later on that the energy of entrons captured by electrons or an ion (the naked proton is an ion) is equivalent to electrical energy on an atomic scale. All entrons have mass which is added to any particles capturing the entron, however the mass of entrons having energies less than about 1 eV is insignificant compared to the mass of a naked electron, and the mass of entrons having energies less than about 1 X 10^5 eV is insignificant compared to the mass of a naked proton. Entron masses, energetic electron masses and energetic proton masses are shown in **Table VI** for entron energies from zero joules (zero eV) to the neutrino entron energy of 1.487 X 10^{-10} joules (9.33 X 10^8 eV). **Table VII** provides energetic electron and proton energies and energetic electron velocities for the same range of entron energies. **Table VII** indicates that it should be easy to accelerate electrons to speeds greater than the speed of light (3 X 10^8 m/s) but it may not be possible to accelerate protons as fast as the speed of light for several reasons. One reason is that the most energetic entron in our Universe (at least in accordance with the current version of the Ross Model) is the neutrino entron with an energy of 1.487 X 10^{-10} joules which results in a speed of 2.98 X 10^8 m/s. Also, in order for entrons to combine with protons, their diameter must be larger than the diameter of the electron orbit in the proton. This limits the individual entron diameters to larger than about 0.85 X 10^{-15} m which corresponds to gamma ray entrons with an energy smaller than 1.02 MeV. However, it may be possible to combine a large number of gamma ray entrons to drive the proton at speeds close to the speed of light or faster up to 1.414 c as explained above.

Table VI
Energetic Electrons and Protons Mass

Entron Energy (Joules) (kg-m^2/s^2)	Entron Energy (eV)	Entron Mass (kg)	Eloctron Mass (kg)	Proton Mass (kg)
zero	zero	zero	9.1094 X 10^{-31}	1.658 X 10^{-27}
1 X 10^{-30}	6.242 X 10^{-12}	1.1126 X 10^{-47}	9.1094 X 10^{-31}	1.658 X 10^{-27}
1 X 10^{-27}	6.242 X 10^{-9}	1.1126 X 10^{-44}	9.1094 X 10^{-31}	1.658 X 10^{-27}
1 X 10^{-24}	6.242 X 10^{-6}	1.1126 X 10^{-41}	9.1094 X 10^{-31}	1.658 X 10^{-27}
1 X 10^{-21}	6.242 X 10^{-3}	1.1126 X 10^{-38}	9.1094 X 10^{-31}	1.658 X 10^{-27}
1 X 10^{-18}	6.242 X 10^{0}	1.1126 X 10^{-35}	9.1095 X 10^{-31}	1.658 X 10^{-27}
1 X 10^{-15}	6.242 X 10^{3}	1.1126 X 10^{-32}	9.2207 X 10^{-31}	1.658 X 10^{-27}
1 X 10^{-12}	6.242 X 10^{6}	1.1126 X 10^{-29}	1.2026 X 10^{-29}	1.68 X 10^{-27}
1.487 X 10^{-10}	9.28 X 10^{8}	1.655 X 10^{-27}	1.655 X 10^{-27}	3.32 X 10^{-27}

Table VII
Energetic Electrons and Protons
Energy and Velocity

Entron Energy (J) $(kg\text{-}m^2/s^2)$	Electron Kinetic Energy (J)	Proton Kinetic Energy (J)	Energetic Electron Velocity (m/s)	Energetic Proton Velocity (m/s)
zero	2.16×10^{-18}	1.34×10^{-12}	$+2.18 \times 10^6$	$+4.02 \times 10^7$
1×10^{-30}	2.16×10^{-18}	1.34×10^{-12}	$+2.18 \times 10^6$	$+4.02 \times 10^7$
1×10^{-27}	2.16×10^{-18}	1.34×10^{-12}	$+2.18 \times 10^6$	$+4.02 \times 10^7$
1×10^{-24}	2.16×10^{-18}	1.34×10^{-12}	$+2.18 \times 10^6$	$+4.02 \times 10^7$
1×10^{-21}	2.16×10^{-18}	1.34×10^{-12}	$+2.18 \times 10^6$	$+4.02 \times 10^7$
1×10^{-19}	2.06×10^{-18}	1.34×10^{-12}	$+2.13 \times 10^6$	$+4.02 \times 10^7$
1×10^{-18}	1.06×10^{-18}	1.34×10^{-12}	$+1.523 \times 10^6$	$+4.02 \times 10^7$
2.18×10^{-18}	zero	1.34×10^{-12}	zero	$+4.02 \times 10^7$
1×10^{-17}	7.84×10^{-18}	1.34×10^{-12}	-3.423×10^6	$+4.02 \times 10^7$
1×10^{-16}	9.8×10^{-17}	1.34×10^{-12}	-2.146×10^7	$+4.02 \times 10^7$
1×10^{-15}	-1×10^{-15}	1.34×10^{-12}	-4.658×10^7	$+4.02 \times 10^7$
1×10^{-13}	$-1. \times 10^{-13}$	1.34×10^{-12}	-4.24×10^8	$+2.73 \times 10^7$
1×10^{-12}	-1×10^{-12}	0.34×10^{-12}	-4.24×10^8	$+2.02 \times 10^7$
1.34×10^{-12}	1.34×10^{-12}	zero	-4.24×10^8	zero
1.49×10^{-10}	1.49×10^{-10}	1.500×10^{-10}	-4.24×10^8	-2.98×10^8

Table VIIA
ELECTRON VELOCITY
With
ENTRON ENERGY IN THE RANGE OF 13.6057 eV

Entron Energy (eV)	Electron Velocity (10^6 m/s)
13.3057	0.459
13.4057	0.265
13.5057	0.180
13.6047	0.059
13.6056	.019
13.60569	0.0018
13.60570	0.0000
13.60571	0.0018
13.6058	0.019

13.6067	0.059
13.7057	0.180
13.8057	0.265
13.9057	0.459

The reader should note from **Table VII** that when the entron energy is zero, we are describing a naked electron and a naked proton. The reader should also note from **Table VII** that when the entron energy is 2.18×10^{-18} J (13.6 eV) the energetic electron's velocity is approximately zero and when the entron energy is 13.4×10^{-13} J (8.37 MeV), the energetic proton's velocity is approximately zero. In a naked proton the electron's speed is increased to 4.72×10^{8} m/s (about 1.57 c) as explained in **Chapter VII** due to additional boost from the electrons own Coulomb force. The two positrons (circling the high energy electron) are traveling at a speed that is about three times faster driven by their own internal Coulomb forces and Coulomb forces from the proton's high energy electron.

Table VIIA shows the velocity of electrons at entron energies required to slow the electron down to velocities close to zero. Note that the velocity units are millions of meters per second. So it probably will be very difficult to reduce the electron's speed to precisely zero.

The most important entron in our Universe is the neutrino entron. We need to know its mass and energy. According to the Ross Model a naked proton is comprised of a naked electron, a neutrino entron and two naked positrons. We have figured out the mass of the naked proton by assuming that the iron-56 atom is comprised of <u>only</u> 56 naked protons and 56 naked electrons, 30 of which are in the nucleus and 26 of which are orbiting or otherwise associated with the iron-56 nucleus, giving the iron atom a net charge of zero.

We know the mass of the iron-56 atom. So we divide the mass of the iron atom by 56 to get the estimated mass of a single naked proton and a single naked electron. (A single naked proton is comprised of one naked electron, one neutrino entron and two naked positrons.) We then subtract the known masses of one naked electron and two naked positrons from the estimated mass of the single naked proton and a single naked electron. The result is the estimated mass of the neutrino entron. Once we have the estimated mass of the neutrino entron, we can estimate its energy using; $E = mc^2$.

We conclude that the neutrino entron has a mass of 1.65496×10^{-27} kg and an energy of 1.48739×10^{-10} J or 9.28×10^8 eV.

CHAPTER X

MASS AND ENERGY OF NEUTRINO ENTRONS

The Most Important

The neutrino entron is the most important subatomic particle in our Universe. We need to know its mass and energy. According to the Ross Model a naked proton is comprised of two positrons and one very energetic electron which is an electron that has captured a neutrino entron. Now we see from **Table II** that the mass of an electron and the mass of a positron is about 9.1093897 X 10^{-31} kg (0.00091094x 10^{-27} kg) each, so the combined mass of an electron and two positrons is about 3 X (0.00091094 X 10^{-27} kg) = 0.0027328 X 10^{-27} kg. So, we can estimate the mass of the neutrino entron if we know the mass of the naked proton. The neutrino entron mass would be the difference between the mass of the naked proton and the combined masses of the two positrons and the electron.

From **Table II** we know that the mass of an ordinary proton (the nucleus of a hydrogen-1 atom) is approximately 1.67262 X 10^{-27} kg. However, we also know that the ordinary proton is significantly more massive that a naked proton because energy is released when ordinary protons are combined to form helium, and additional energy is released when the helium nuclei are combined to form even heaver atoms. In fact, energy is released in all combinations to form heaver atoms from the combination of lighter atoms until we get to iron-56. For atoms heavier than iron-56, energy/mass has to be added in the form of gamma ray entrons to form the heavier atoms from lighter atoms. This version of the Ross Model assumes that the difference in mass between an ordinary proton and a naked proton is gamma ray entrons that have been captured by the naked proton to slow it down (from its natural speed of 4.02 X 10^7 m/s) enough so it can capture an orbital electron to become a hydrogen-1 atom. These captured gamma ray entrons are released in fusion processes when heavier atomic nuclei are formed from the lighter atomic nuclei.

As explained above, I have made the further assumption that iron-56 nuclei contain no significant mass of entrons. My assumption therefore is that the iron 56 atom contains 56 naked protons and 30 naked electrons in its nucleus (to give the nucleus a positive charge of plus 26) and 26 naked electrons in orbit around the nucleus or otherwise associated with the nucleus (to give the atom a net zero charge). We know that the iron-56 atom has a total mass of about 92.8822 X 10^{-27} kg. So we can divide 92.8822 X 10^{-27} kg by 56 to determine the mass of a single naked proton and one naked electron. This means, based on these assumptions, that a single naked proton and one naked electron would have a mass of 1.65861 X 10^{-27} kg. The naked electron has a mass of 0.00091044 X 10^{-27} kg, so the mass of the naked proton is estimated to be: $m_p = 1.6576956 \times 10^{-27}$ kg.

Now we subtract the masses of the naked electron and two naked positrons (i.e. $0.0027328 \times 10^{-27}$ kg) from our estimate of the mass of the naked proton (i.e. 1.65861×10^{-27} kg) to obtain the mass of the neutrino entron:

$$m_{\text{neutrino entron}} = 1.6576956 \times 10^{-27} \text{ kg} - 0.0027328 \times 10^{-27} \text{ kg}$$
$$= 1.65496 \times 10^{-27} \text{ kg}.$$

We get the energy of the neutrino entron using Professor Einstein's famous formula:

$$E = mc^2$$
$$E_{\text{neutrino entron}} = (1.65496 \times 10^{-27} \text{kg}) (2.99792 \times 10^8 \text{ m/s})^2$$
$$E = 1.48739 \times 10^{-10} \text{ J}$$

The energy in electron volts is:

$$E_{\text{neutrino entron}} = (1.48739 \times 10^{-10} \text{ J})/1.602 \times 10^{-19} \text{ J/eV}$$
$$= 928 \times 10^6 \text{ eV} = 928 \text{ MeV} = 9.28 \times 10^8 \text{ eV}$$

The neutrino photon has the same energy and mass as the neutrino entron.

The reader should take another look at **Table V** in **Chapter V** to see how the neutrino entron and its neutrino photon compare to other entrons and photons.

Crab Nebula

Image credit: ESA/Herschel/PACS/MESS Key Programme Supernova Remnant Team; NASA, ESA and Allison Loll/Jeff Hester (Arizona State University)

Scientists know a lot about photons; but they are not aware of the structure of the photons. As of the writing of this book, they know nothing about entrons. The Ross Model proposes that tronnies travel in entrons in the form of a circle, that entrons travel in photons in the form of a circle, and that the energy of a photon is equal to the energy of its entron. The model also proposes that the diameter d of the photon's circle is related to the photon's wavelength λ by: $d = 2\lambda / \lambda = 0.6366\lambda$. The Ross Model proposes that all entrons and photons have the same structure. So if we know the ratio of the diameter of a single photon's circle to the diameter of its corresponding entron's circle, we will know the corresponding ratios for all photons to their corresponding entrons.

The Ross Model uses the neutrino entron to calculate the ratio of photon diameters d to entron diameters d':

$$d/d' = 911.$$

84

CHAPTER XI

SIZES OF NEUTRINO ENTRONS AND NEUTRINO PHOTONS

The Size of the Neutrino Entron

The neutrino entron is the most massive entron in our Universe. It is also the smallest entron in our Universe. In fact, aside from tronnies that have no size at all, the neutrino entron is the smallest thing in our Universe. It is about one-half the size of a naked electron or a naked positron which are tied for the second smallest things in our Universe (again excluding tronnies). Neutrino entrons are so small that nearly all of them in a beam of neutrino photons can pass through the earth, giant planets and stars at the speed of light with no problem. How small are neutrino entrons? Here is where we estimate the size of the neutrino entron.

Each neutrino entron, like all entrons, is comprised of one plus tronnie and one minus tronnie as shown in **FIG. 2A**. Keep in mind that the neutrino entron (like all other entrons) has no thickness since each of its two tronnies has no size, but the neutrino entron, like all entrons, does have a diameter and a circumference. We will define the size of our entrons by their diameter d'. Keep in mind that photons, including neutrino photons also have a diameter as we have explained above and also no thickness. We will refer to the diameter of photons as d. In this section we will calculate our estimate of the ratio of d to d' for the neutrino photon and the neutrino entron. Once we have the ratio, we will use that ratio for all entrons and their respective photons.

Diameter d of the Neutrino Photon

First we need to calculate the diameter d, of the neutrino photon. Since we have estimated in **Chapter X** the energy E of the neutrino entron as 9.28×10^8 eV and since we assume that the energy of the neutrino photon is the same as the energy of its entron, we can calculate the wavelength λ and the diameter d of the neutrino photon using equations (1) from **Chapter I** and (4) from **Chapter V**:

$$E = hc/\lambda \qquad (1)$$
$$\lambda = hc/E \text{ and}$$
$$d = 0.6366\lambda \qquad (4)$$

From the above equations we see that:

$$\lambda_{\text{neturino photon}} = hc/E = (4.135 \text{ X } 10^{-15} \text{eVs})(2.9979 \text{ X } 10^8 \text{ m/s})/9.28 \text{ X } 10^8 \text{ eV}$$
$$= 1.336 \text{ X } 10^{-15} \text{ m}$$

and that:

$$d_{\text{neutrino photon}} = (0.6366)1.336 \text{ X } 10^{-15} = 0.85 \text{ X } 10^{-15} \text{ m}.$$

We use the same equations to estimate the diameter of the 1.02 MeV photon. Since the wavelength of the 1.02 MeV gamma ray photon is $\lambda = hc/E = (4.135 \text{ X } 10^{-15}$ eVs)(2.99X10^8 m/s)/1.02 X 10^6eV which is about 1.213 X 10^{-12} m, so the diameter d of the circle the 1.02 MeV gamma ray photon from **Equation (4)** is assumed to be:

$$d_{1.02 \text{ MeV GR Photon}} = 0.6366\lambda = (0.6366)(1.213 \text{ X } 10^{-12} \text{ m}) = \text{about } 7.722 \text{ X } 10^{-13} \text{ m}.$$

The Diameter of the Neutrino Entron and the Ratio d/d'

In order to understand the concepts developed in this section the reader should be familiar with the currently understood process of "electron – positron pair production" and "electron-positron annihilation". If you are not, you may want to GOOGLE these phrases. Existing scientific explanations are referred to briefly in **Chapter I**. The Ross Model provides an explanation of these processes that is substantially different for existing scientific explanations.

The Ross Model description of pair production is one in which:
* the two tronnies of a 928 MeV neutrino entron (where d' = 0.9339 X 10^{-18} and d = 0.8499 X 10^{-15} m),
* the two tronnies of a 1.0214 MeV gamma ray entron (where d' = 0.8485 X 10^{-15} and d = 7.722 X 10^{-13} m) and
* the two tronnies of a 1.122 KeV ultraviolet entron (where d' =7.722 X 10^{-13} and d = 7.028 X 10^{-10} m);

(totalling 3 plus tronnies and 3 minus tronnies) combine to form an electron and a positron (together having a total of 3 plus tronnies and 3 minus tronnies).

We know that pair production takes place only when gamma radiation at energies equal to or greater than 1.02 MeV interacts with matter. We also know that a 1.02 MeV gamma ray photon has a wavelength of about $\lambda = 1.22$ X 10^{-12} m and that the size of electrons and positrons are almost a million times smaller at about 2 X 10^{-18} m. The Ross Model assumes that the three entrons listed above are resonant with each other and that they are resonant with each other because:

1) the diameter of the 928 Mev neutrino entron is equal to about 0.9339×10^{-18} m,

2) the photon circle of the 928 MeV neutrino entron has a diameter of about 0.85×10^{-15} m approximately of equal to the 0.8485×10^{-15} m diameter of the 1.02 MeV gamma ray entron, and the photon of the 1.02 MeV entron has a diameter of about 7.722×10^{-13} m, and

3) the entron of the 1.122 KeV ultraviolet photon has a diameter of about 7.722×10^{-13} m approximately equal to the 7.722×10^{-13} m diameter of the 1.02 MeV entron.

We know that gamma ray photons including the 1.02 MeV gamma ray photon come from the nuclei of atoms which have diameters in the range of 10^{-15} m to 10^{-14} m, (i.e. much smaller than the wavelength and the diameter of the gamma ray photons). According to the Ross Model, this is possible because the thing that is a gamma ray photon on the outside of an atomic nuclei is a gamma ray entron on the inside of the nuclei. The question is how much smaller is the diameter of the entron as compared to the diameter of its photon?

We know that the energy of the 1.02 MeV gamma ray photon is 1.02 MeV. We assume that this is also the energy of its entron. Now we have estimated in **Chapter X** that the energy of the neutrino entron is 928 MeV. This estimate was derived from our assumption that the naked proton is comprised of a naked electron, a neutrino entron and two naked positrons and that the neutrino entron provides all of the mass of a naked proton except for the relative small masses of the naked electron and the two positrons. We have also assumed that the energy of each entron is inversely proportional to the diameter of the entron, and we have further assumed that the wavelength and the diameter of a photon are proportional to its entron's diameter. Now the question is how could both the gamma ray entron and the neutrino entron be resonate with the plus tronnie of a naked electron? The answer has been suggested above and the answer is that each entron defines two circles:

(1) the circle formed by its two tronnies defining a diameter d' (see **FIG. 2A**) of the entron, and

(2) the circle formed by the entron defining a diameter d when it travels as a photon (see **FIG. 3**).

Therefore the Ross Model proposes that the neutrino photon is resonate with the naked electron because the diameter d' of its entron is equal to the diameter of the circle defined by path 300 of the plus tronnie in the naked electron (see **FIG. 5** in **Chapter VII**) and that neutrino entrons being resonant with naked electrons can be captured by naked electrons. The Ross Model also proposes that once a neutrino entron with a mass-energy

87

equal to 1.654596 X 10^{-27} kg (1.48739 X 10^{-10} J = 928 MeV) is captured by an electron with the much smaller mass-energy of 0.00091 X 10^{-27} kg (0.51 MeV), the electron is driven along the much larger circular path of the neutrino entron in the neutrino photon. Next we assume that the diameter of the 1.02 MeV gamma ray entron is equal to the diameter of the neutrino photon. We have calculated the diameter of the neutrino photon in the section above entitled "Diameter d of the Neutrino Photon" based on the estimate we made of the energy-mass of the neutrino photon which was based on our assumption that almost all of the mass of a naked proton is attributable to a neutrino entron captured by the naked electron in the naked proton. Our estimate of the diameter of the neutrino photon is $d_{neutrino\ photon}$ = 0.85 X 10^{-15} m. So our assumption is that the diameter d' of the 1.02 MeV gamma ray entron is equal to the diameter d of the neutrino photon, i.e.:

$$d'_{1.02\ MeV\ GRentron} = d_{neutrino\ photon} = 0.85\ X\ 10^{-15}\ m.$$

Since we have determined that the diameter of the 1.02 MeV entron is 7.77 X 10^{-10} m, we can estimate the ratio of the diameter d of the 1.02 MeV photon to the diameter d' of its 1.02 MeV entron of:

$$d/d' = (7.77\ X\ 10^{-13}\ m)/(0.85\ X\ 10^{-15}\ m) = 911\ and$$
$$d/d' = 911. \qquad (5)$$

This ratio d/d' = 911 for the 1.02 MeV gamma ray entron is very important to the Ross Model since according to the model all entrons are exactly alike except for size, mass and energy, all of which define an electro-magnetic spectrum of about 17 orders of magnitude. So if we know d/d' for one photon-entron combination, we know it for all photon-entron combinations. So let us apply this knowledge first to estimate the diameter of the neutrino entron. Thus:

$$d'_{neutrino\ entron} = (d_{neutrino\ photon})/911 = (0.85\ X\ 10^{-15}\ m)/911$$
$$= 0.934\ X\ 10^{-18}\ m.$$

So this is the diameter of a neutrino entron according to the Ross Model. If you want you can round this number off to d' = about 1 X 10^{-18} m to make it easier to remember or we could try to remember that the entron diameter is about one thousand times smaller than the diameter of the neutrino photon.

Size of the Electron

Based on the assumptions we were making in determining the size of the neutrino entron we can now estimate the size of the naked electron. Our assumption was that the diameter of the neutrino entron is equal to the diameter of the circular path 300 of the plus tronnie in the naked electron. (Take a look at **FIG. 5** in **Chapter VII.**) So this would mean that the diameter of the plus tronnies' circle is about 1×10^{-18} m. The diameter of the circles of the two minus tronnies in the electron is equal to the diameter of the plus tronnies circle. The paths of the two minus tronnies are looping through the center of the plus tronnies' circle giving the electron the shape shown in **FIG. 5**. So my estimate to the size of the naked electron is a little less than 2×10^{-18} m.

Protons, electrons and entrons - and only protons electrons and entrons - combine to make neutrons, deuterons, tritons and alpha particles. (See FIGS. 9 – 11.)

Quarks do not exist. Neutrons exist only for a few minutes; then they decay to an electron, proton and a gamma ray entron-photon, whether they are inside or outside an atomic nucleus.

CHAPTER XII

NEUTRONS, DEUTERONS, TRITONS AND ALPHA PARTICLES

Neutrons

The Ross Model proposes a new model of the neutron. According to the Ross Model the neutron is made of a naked proton and a naked electron with a combined mass of 0.99883825 atomic mass units (amu) and a number of high-energy entrons sufficient to provide the neutron with additional mass so that its measured mass is 1.00866492 amu. The difference is 0.00982667 amu, equivalent in energy to 9.1585 MeV and in regular mass units to 1.6376×10^{-29} kg. (See **Table XII** in **Chapter XIII.**) According to the Ross Model a neutron is a very short lived particle with an average lifetime of about 15 minutes whether it is inside or outside a nucleus of an atom. In both cases it quickly decays into its constituent parts. The parts of the neutron (the electron, the proton and the high-energy entrons) exist and function inside the nucleus of an atom as separate components of the nucleus.

Two or three combinations of a proton, an electron and gamma ray entrons are released together in the form of a neutron in the process of the fissioning of certain nuclei such as uranium-235 and plutonium-239. In the fission process (in atomic bombs and nuclear reactors) the velocity of these neutrons is about 2.8×10^7 m/s (about 9 percent of the speed of light, representing a kinetic energy of about 4.09 MeV.). They rightly are called "fast neutrons". For a comparison, according to the Ross Model the estimated velocity of the naked proton is a little faster at about 4.0226×10^7 m/s. According to the Ross Model naked neutrons like naked electrons and naked protons are self-propelled and like the naked electrons and protons, slow down by capturing high-energy entrons. (Most likely the fast neutrons speed is attributable to the proton portion of the neutron.) In most nuclear reactors, water is provided to help slow down the neutrons. Scientists believe the neutron loses its kinetic energy by inelastic collisions (similarly to billiard ball collisions) with the hydrogen nuclei in the water. According to the Ross Model the neutron does not lose its velocity by collisions; instead, it loses its natural self-propelled velocity by stealing gamma ray entrons from water molecules and other particles in its environment, and these stolen entrons once captured by the neutron reduce the natural speed of the neutron. (Remember from **Chapter V** low-speed protons in hydrogen atoms owe their low speed to a collection of several gamma ray entrons having a total energy of about 8.37 MeV. Also, all other atoms except possibly iron-56 also have extra gamma ray entrons within their nuclei.) In a nuclear reactor fast neutrons are quickly slowed down to a few thousand meters per second depending on the temperature of their environment and its energy at these speeds is referred to as "thermal energy". At about 300 K (80.6 °F and 27 °C) the speed of the thermal neutrons would be about 4.9 X 10³ m/s, 4,900 meters per second, still pretty fast. This gives the thermal energy neutron (with a mass of about 1.67 X 10⁻²⁷ kg) a kinetic energy of about 20.05 X 10⁻²¹ J (equivalent to about 0.126 eV).

Often a neutron will be absorbed in the nuclei of another atom before it decays into a proton and an electron in which case the components of the neutron becomes a part of the absorbing nuclei. In many cases the nuclei absorbing the neutron will be radioactive as a result of the absorption and as a result will release an electron or a proton or an entron (in the form of a gamma ray photon) in the course of returning to a stable condition. This is further support for the Ross Model contention that a neutron is merely a combination of a proton, an electron and one or more entrons.

Outside the nucleus if the neutron is not absorbed quickly it will decay with an average lifetime of about 15 minutes. When the neutron decays after this brief lifetime, the decay products are the proton and an electron. (The electron referred to as a beta particle has an energy of 1.253 X 10⁻¹³ J corresponding to 13.9 X 10⁻³¹ kg. According to the Ross Model this is the energy of the entron of the beta particle, E_e. If we use **Equation (9)** from **Chapter IX** and this energy of the electron we can estimate the velocity of the electron. The natural energy of the electron E_N is about 10 thousand times smaller than the energy E_e of the entron, so E_N can be ignored and $E_N - E_e = E_e$, therefore:

$$v = \sqrt{\frac{2(E_N - E_e)}{m_N + m_e}} = \sqrt{\frac{2(1.253\ X\ 10^{-13}J)}{9.109\ X\ 10^{-31}kg + 13.9\ X\ 10^{-31}kg}} = \sqrt{10.89\ X\ 10^{16}m^2/s^2}$$

$$= 3.3\ X\ 10^8 m/s$$

(The reader should notice here that the Ross Model predicts that the electron in the neutron is traveling about **10 percent faster than the speed of light!** I know this is going to disturb a lot of people since most people believe what Professor Einstein told us: that nothing can go faster than the speed of light. The reader should also note however that the electron mass has increased from 9.109 X 10⁻³¹ kg to 23.009 X 10⁻³¹ kg. However, the increase in mass is not the result of the electron going fast; the increase is the result of the capture by the electron of the 13.9 X 10⁻³¹ kg entron.)

We can now estimate the orbit radius of the electron in the neutron by applying Newton's second law, which asserts that the attractive force on the electron from the proton is equal to the electron mass times the radial acceleration of the electron so:

Attractive Force = Electron Mass times Radial Acceleration

$$\frac{ke^2}{r^2} = \frac{mv^2}{r}$$

So:

$$r = \frac{ke^2}{mv^2} = \frac{(8.99X10^9 Nm^2/C^2)(1.602X10^{-19}C)^2}{(23.009X10^{-31}kg)(3.3X10^8 m/s)^2} = 0.92\ X\ 10^{-15}m$$

The radius of 0.92 X 10^{-15} m is slightly larger than the Ross Model's estimate of the 0.85 X 10^{-15} m diameter of the high-energy electron's path in the naked proton. (See **FIG. 7** in **Chapter VIII.**) Based on these calculations my model of the neutron is a naked proton with a 0.782 MeV energetic electron circling through or around the proton. The reader should note that the diameter of the entron of the 0.782 MeV photon according to **Equation 10** (d' = 8.677 X 10^{-10} eVm/Ep) is d' = about 1.1 X 10^{-15} m, also somewhat larger than the estimated size of the naked proton. So I propose that the electron in the neutron is driven by its entron along the circular path of its entron, passing through the center of the proton each cycle.

The Deuteron (Deuterium Nucleus)

Deuterium is an isotope of hydrogen. Its natural abundance is a very small percentage of the abundance of hydrogen-1 in nature. It is almost twice as massive as hydrogen-1 (see **Table XI**). Current physics and chemistry books explain that the deuterium nucleus is comprised of a proton and a neutron plus some binding energy. The deuterium nucleus (an energetic deuteron) according to the Ross Model is simply two naked protons plus a single naked electron and a number of entrons having a total mass/energy equivalent to 15.29 MeV. A 15.29 MeV entron is a gamma ray entron with a diameter of about 0.890 X 10^{-16} m. A 0.89 X 10^{-16} m entron diameter is smaller than the diameter of a proton; therefore, the Ross Model proposes that several lower energy gamma ray entrons (with larger diameters) are probably utilized to slow down the deuterium nucleus. And the Ross Model proposes that the naked deuteron is self-propelled with a kinetic energy of 15.29 MeV. A proposed drawing of the naked deuteron is shown in **FIG. 9**. The gamma ray entrons are not shown. Many of these gamma ray entrons are released when the deuteron is fused in hydrogen bombs to form helium.

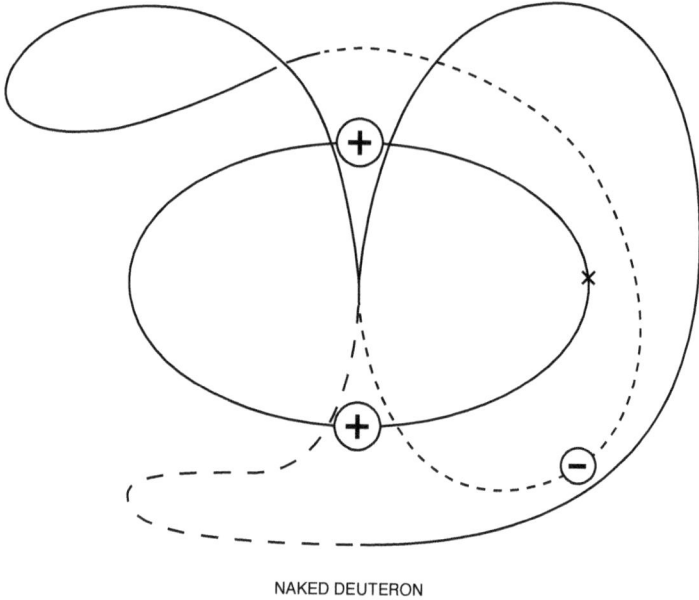

NAKED DEUTERON

FIG. 9

The Triton (Tritium Nucleus)

Tritium is another isotope of hydrogen, almost three times as massive as hydrogen-1. Current physics and chemistry books explain that the tritium nucleus (a triton) is comprised of a proton and two neutrons plus some binding energy. The tritium nucleus according to the Ross Model is simply three naked protons plus two naked electrons and enough gamma ray entrons to provide an energy/mass of about 18.187 MeV. The Ross Model proposes that the naked triton is self-propelled with a kinetic energy of 18.187 MeV. Several entrons are probably required to provide the breaking action. A proposed drawing of the triton is shown in **FIG. 10**. The entrons are not shown. As with the deuteron when the triton is utilized in fusion bombs at least some of these gamma ray entrons are released as destructive energy. The tritium nucleus is radioactive and decays with a 12.32 year half-life with the release of a 0.01859 MeV beta particle (an electron) to produce a helium-3 nucleus which is a rare stable isotope of helium.

It is interesting to note that the helium-3 nucleus with three protons and only one electron is stable (i.e. unlimited half-life) whereas the tritium nucleus with three protons and two electrons has a limited 12.32 year half-life. You might think three plus charges and two

94

minus charges (tritium) would be more stable than three plus charges and one minus charge (helium-3), but in this example that is not the case. However, if this situation is viewed under the Ross Model, we see that each of the three protons is comprised of two low-mass plus charges and one relatively very large mass minus charge. So the helium-3 nucleus can be stable with 3 large-mass minus charges, 6 low-mass positive charges and 1 low-mass minus charge for a net charge of plus 2; whereas tritium (H-3) is unstable with 3 large-mass minus charges, 6 low-mass plus charges and 2 low-mass minus charges with a net charge of plus 1.

The Alpha Particle (a Helium Nucleus)

Current physics and chemistry books explain that the alpha particle (which is the same as the helium-4 nucleus) is comprised of two protons and two neutrons plus some binding energy. (Almost 100 percent of helium isotopes are helium-4 isotopes.)

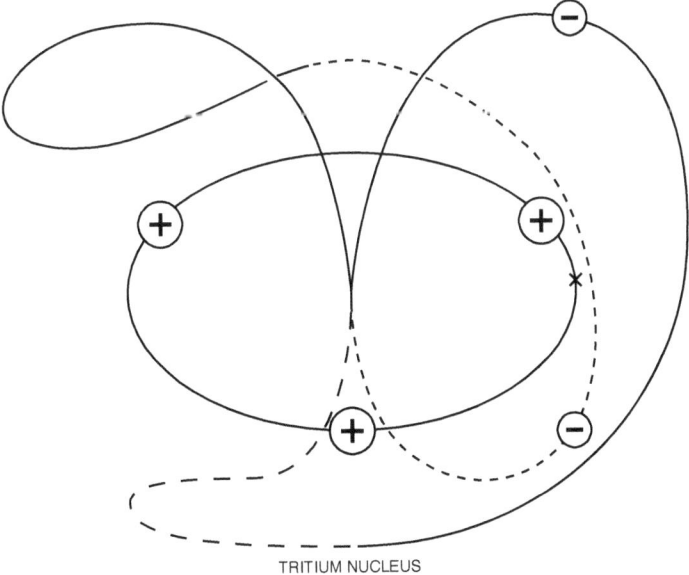

TRITIUM NUCLEUS

FIG. 10

95

The alpha particle according to the Ross Model is simply four naked protons plus two naked electrons and a number of entrons having a total mass/energy equivalent to about 6.75 MeV (see **Table XII**). A proposed drawing of the alpha particle is shown in **FIG. 11**. Again, the entrons are not shown. The reader should take notice of the structure of the deuteron, the tritium nucleus and the alpha particle, for these particles will be used by the physics of stars, according to the Ross Model, to build larger atoms (without the magical "strong force"). Notice in each case the particles comprise a negative portion and a positive portion but the net charge of each of the nuclei is positive. However, the positive portion in each case is located in the center of the particle. The outside portion of each of the three particles is negative. Thus the negative portion of each alpha particle is much closer to its neighbor than the positive portion. Also in the alpha particle there are really 14 charges, 8 plus charges and only 6 minus charges so it has a net charge of plus 2, so you might think all alpha particles would repel other alpha particles. However, the Ross Model teaches that 4 of the minus charges are each about 1,000 times heavier than each of the plus charges. The net effect is that at very close quarters alpha particles can be attractive to other alpha particles. We will see in the next chapter that this attractiveness of alpha particles to each other at short ranges is extremely important in the process of building larger atoms. It is very important to remember that Coulomb forces between stationary charges decrease inversely proportionally to the square of the distance between charges. We will see in the next chapter, **Chapter XIII**, that the nuclei of one particular isotope of carbon, oxygen, neon, magnesium, silicon, sulfur, argon and calcium are comprised of nothing but alpha particles (see **Table 10**) and entrons. When the alpha particles are very close together, but not too close, the Coulomb attractive forces between the negative and positive portions of adjacent alpha particles are strong enough to exactly balance the repulsive forces acting between the particles. For all of these atoms other than argon, the particular isotope formed entirely of alpha particles plus entrons is the most abundant naturally occurring isotope of the atom. The construction of these heavier atoms is discussed in the next chapter, **Chapter XIII**.

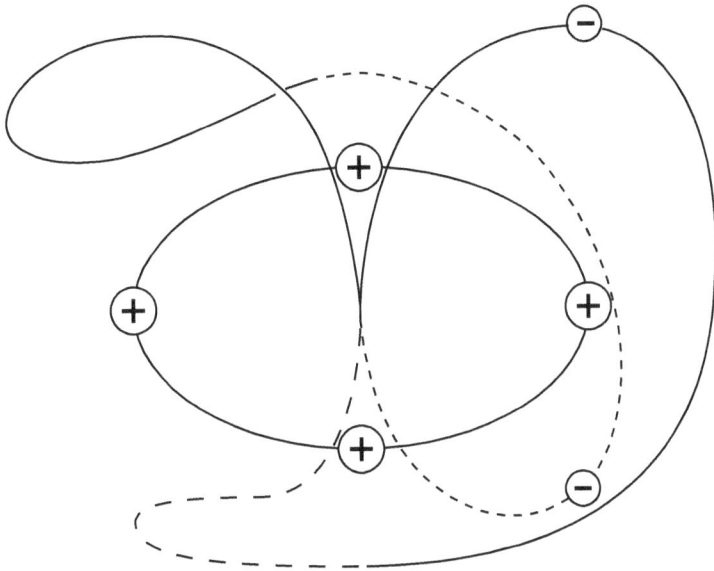

NAKED ALPHA PARTICLE

FIG. 11

Most stable atomic nuclei are comprised only of alpha particles and entrons or alpha particles, electrons and entrons.

An alpha particle is four circling protons and two electrons circling through the circular path of the circling protons. (See FIG. 1 in Chapter XII.) The four positively charged protons in alpha particles are attracted to the negatively charged electrons of close-by alpha particles so close-by alpha particles are attractive to each other in specific configurations, but not all configurations. For example two close-by alpha particles form a very unstable beryllium-4 nucleus, but three close-by alpha particles form a very stable carbon-12 nucleus. No "strong force" is needed. The Coulomb forces are sufficient. There is no such thing as the so-called "strong force" holding atomic nuclei together.

Similarly, the most abundant isotopes of oxygen, neon, magnesium silicon, sulfur and calcium are all comprised of nothing but alpha particles and some gamma ray entrons. Nuclei heavier than calcium need some additional high-energy electrons in their nuclei for stability.

CHAPTER XIII

HOW TO MAKE ATOMS

Building Blocks for Atoms

As I have stated many times, everything in our Universe is made from nothing but tronnies or things made from tronnies. As we have seen above, tronnies can be combined to make entrons, electrons and positrons; and entrons, electrons and positions can be combined to make protons. Protons, electrons and entrons can be combined to make deuterons and alpha particles. Now we use these composite particles (deuterons and alpha particles) along with electrons, positrons and protons to make the light atoms up through iron-56. And then the nuclei of light atoms along with entrons and electrons can be combined to make heaver atoms up through uranium 238. And all of the atoms can be combined in an almost unlimited way to make molecules; and entrons, atoms and molecules can be combined to make everything else in our Universe including us.

The Sizes of Atoms

The sizes of atoms are generally in the range of about 10^{-10} meters. Almost all of the mass of each atom is contained in its nucleus which is about 1000 times smaller than the atom. A number of electrons between 1 and 92 circle the nucleus (or are otherwise associated with the nucleus) in the 92 naturally occurring atoms. The sizes of nuclei can be approximated by the following formula:

$$r = r_0 A^{1/3}$$

where r is the radius of the nucleus, $r_0 = 1.2 \times 10^{-15}$ m and A is the mass number of the isotope. So, for example, iron-56 (with a mass number of 56) has a nucleus with a radius of about 5.5×10^{-15} m, since the cube root of 56 is about 4.59. The largest stable isotope in our Universe is uranium-238. Its nucleus has a radius of about 6.2×10^{-15} m. Helium with a mass number of 4 has a radius of about 1.92×10^{-15}. Oxygen has a mass number of 16, so its nucleus has a radius of about 3×10^{-15} m.

Making Models of Atoms

The question for now is how are these atoms and molecules assembled? To help figure this out, first we will consider the atomic number and masses of atoms. Most atoms have several naturally occurring isotopes. The atomic number of all isotopes of all atoms is equal to the number of electrons associated with the nucleus of the charge neutral isotope (i.e. 1 for the isotopes of hydrogen and 92 for the isotopes of uranium). For example, all charge neutral oxygen isotopes have eight associated electrons, so all isotopes of oxygen have an atomic number of eight.

Molecular Hydrogen

Hydrogen exists at normal temperatures as a molecule which is a combination of two atoms of hydrogen. This molecule is referred to as H_2. This molecule can be broken apart but it takes energy to do it. The energy required is 145.6 kilojoules per mol of H_2. A kilojoule per mol is equivalent to 1.0365×10^{-2} eV per atom or molecule as explained in **Table IV**. So it takes about 1.5 eV of energy to split a hydrogen molecule into its two parts. This can be provided by two 0.75 eV photons in which each of the two atoms would leave with an entron with energy of about 0.75 eV.

According to the Ross Model an H_2 hydrogen molecule consists of two low-speed, high-energy hydrogen atoms in their ground state. Remember from **Chapter VIII** the nucleus of each hydrogen atom consists of a naked proton which has captured several gamma ray entrons (about 15) having a total energy of about 8.37 MeV which have caused the naked proton to slow down from about 4.03×10^7 m/s to a speed relatively close to zero speed. But keep in mind the speed of the low-speed proton is not zero. On the average, the speeds are probable a small fraction of the speed of the naked proton. For example if the gamma ray entrons have reduced the speed of the proton by a factor of 1,000, its speed would be in the range of about 4×10^5 m/s, 40 thousand meters per second. These two positively charged protons probably exist together in a structure in which the two protons circle on a common circle and the two electrons circle through the center of the circle.

Atomic Hydrogen

All but about 1.5 percent of hydrogen atoms consist of a single proton (see **Chapter VIII**) and an associated electron. Another isotope of hydrogen is deuterium with two protons and an electron in the nucleus and one electron associated with the nucleus. A third isotope is tritium with three protons and two electrons in the nucleus and one additional electron associated with the nucleus. Tritium is radioactive and decays with an emission of an electron (with a half-life of 12.32 years) to form a rare isotope of helium (helium-3).

All Other Atoms

According to the Ross Model the nuclei of all stable atoms are comprised of only electrons, protons, and high-energy entrons or particles made of electrons, protons and high-energy entrons. (Keep in mind that each proton is a combination of a massive electron which is an electron combined with a neutrino entron, two positrons and some additional high-energy entrons.) Some of these electrons, protons and high-energy entrons are combined in composite particles (i.e. deuterons, tritons and alpha particles). A low-energy entron cannot exist in an atomic nucleus, simply because it is much too large. All entrons in nuclei (other than the neutrino entron in each proton) are gamma ray entrons. Some of these high-energy entrons have converted the composite particles into energetic composite particles. We have estimated the mass of the naked proton, we know the mass of the naked electron. The actual mass of all isotopes of all atoms are known with great accuracy from careful measurements.

Building of Atoms

According to the Ross Model, the nuclei of most of the isotopes of most atoms are comprised mostly of alpha particles, the nuclei of helium atoms. Atoms are made in stars. There are lots of alpha particles in most stars because most stars are mostly made of hydrogen and helium. However, at the temperature of the internal structure of stars the hydrogen and helium atoms are ionized which means that we are talking about the nuclei of helium and hydrogen (i.e. alpha particles and protons) and separate electrons. Stars are extremely hot and that "hot" is nothing but high-energy entrons as we have explained above, so there is an enormous quantity of high-energy entrons available in stars. And at the temperature at the core of stars the entrons are gamma ray entrons. According to the Ross Model these composite particles are the building blocks of the nuclei of atoms.

In **Table X**, I have listed 61 isotopes having a spin of zero and a mass number that can be evenly divided by four. The ten lightest of these **Table X** isotopes are believed by me to be comprised of nothing but alpha particles and entrons. I am proposing that the heavier isotopes listed in **Table X** also contain, in addition to the alpha particles and entrons, the number of electrons specified in **Table X**.

In **Table XI**, I have listed some typical isotopes of typical atoms along with their atomic masses and spins. These nuclei can be constructed with alpha particles and entrons plus an appropriate number of additional electrons and/or protons which may be in the form of deuterons.

Table X
Alpha Particle Combinations to Form Zero Spin Atomic Nuclei

Isotope	Atomic Number	Mass Number	Number of Alpha Particles	Electrons	Percent Abundance
Helium 4	2	4	1	0	100
Beryllium 8	4	8	2	0	0*
Carbon 12	6	12	3	0	98.9
Oxygen 16	8	16	4	0	99.8
Neon 20	10	20	5	0	90.5

Magnesium 24	12	24	6	0	79.0
Silicon 28	14	28	7	0	92.2
Sulfur 32	16	32	8	0	95.0
Argon 36	18	36	9	0	0.336
Calcium 40	20	40	10	0	96.9
Argon 40	18	40	10	2	99.6
Calcium 44	20	44	11	2	2.1
Titanium 48	22	48	12	2	73.7
Chromium52	24	52	13	2	83.8
Iron 56	26	56	14	2	91.8
Nickel 60	28	60	15	2	26.2
Zinc 64	30	64	16	2	48.6
Zinc 68	30	68	17	4	18.4
Germanium 72	32	72	18	4	27.7
Germanium 76	32	76	19	4	7.44
Selenium 80	34	80	20	6	46.6
Krypton 80	36	80	20	4	2.25
Krypton 84	36	84	21	6	57.0
Strontium 84	38	84	21	4	0.56
Strontium 88	38	88	22	6	82.6
Zirconium 92	40	92	23	6	17.1
Zirconium 96	42	96	24	8	2.8
Molybdenum 100	42	100	25	8	9.36
Rubidium 104	44	104	26	8	18.7
Palladium 108	46	108	27	8	26.5
Cadmium 112	48	112	28	8	24.1
Cadmium 116	48	116	29	10	7.5
Tin 120	50	120	30	12	32.6
Tin 124	50	124	31	12	5.8
Tellurium 128	52	128	32	12	31.7
Xenon 132	54	132	33	12	26.9
Xenon 136	54	136	34	14	8.9
Barium 136	56	136	34	12	7.8
Cesium 140	58	140	35	12	88.5
Neodymium 144	60	144	36	12	23.8
Neodymium 148	60	148	37	14	5.7
Samarium 148	62	148	37	12	11.3**
Samarium 152	62	152	38	14	26.7
Gadolinium 156	64	156	39	14	20.5

Gadolinium 160	64	160	40	16	21.9
Dysprosium 160	66	160	40	16	2.34
Dysprosium 164	66	164	41	16	28.2
Erbium 164	68	164	41	14	1.61
Erbium168	68	168	42	14	26.8
Ytterbium 172	70	172	43	16	21.9
Ytterbium176	70	176	44	18	12.9
Tungsten 180	74	180	45	16	0.12
Tungsten 184	74	184	46	18	30.6
Osmium 188	76	188	47	18	13.3
Osmium 192	76	192	48	20	41.0
Platinum 196	78	196	49	20	25.3
Mercury 200	80	200	50	20	23.1
Mercury 204	80	204	51	22	6.87
Lead 204	82	204	51	20	1.4
Lead 208	82	208	52	22	52.4

There are no naturally occurring 53-57 alpha isotopes.

Thorium 232	90	232	58	26	100***
Uranium 236	92	236	59	26	0****

* Beryllium-8 half-life is 7 X 10^{-17} seconds – decay products are two alpha particles.

**Samarium-62 half-life is 7 X 10^{15} years - decay product is an alpha particle.

***Thorium-232 half-life is 1.4 X 10^{10} years - decay product is an alpha particle.

**** Uranium 236 half-life is 2.34 X 10^7 years - decay product is an alpha particle.

We know that atoms have spin and we know the spin is not the same in all atoms. So we can now utilize all of this information to predict how each of the atoms is constructed. I have not tried to describe all isotopes of all atoms. There are 92 distinct naturally occurring types of atoms, each type having its own atomic number and chemical properties. Most of these 92 atoms have more than one stable isotope and many unstable isotopes. So there are several hundred isotopes. I have picked only a few isotopes for examples and will try to describe their structure. Information for **Tables X, XI** and **XII** are extracted from the 77th (1996-1997) Edition of the CRC Handbook of Chemistry and Physics, published by CRC Press.

In **Table XII**, I have listed the calculated mass difference between the measured masses of some typical isotopes and the sum of the Ross Model estimated masses of the naked protons and electrons making up each of the listed isotopes. I have also included the electron and the neutron. This difference, between (1) the sum of the masses of the naked electrons and the naked protons in the isotopes and (2) the measured masses of the isotopes, according to the Ross Model, is equal to the mass of the entron or entrons in the isotope. You should notice that this mass difference for the iron-56 isotope is zero. This results from the fact that I used this isotope to estimate the mass of the naked proton under the assumption that there were no entrons with significant mass in the iron-56 isotope.

Table XI
Atomic Masses and Spin

Atom	Atomic Number	Atomic Isotope	Abundance (percent)	Atomic Mass (amu)	Spin
Electron				0.00054856	1/2+
Neutron	0	^{0}n		1.00866492	1/2+
Hydrogen	1	^{1}H	99.985	1.00782	1/2+
Deuterium	1	^{2}H	0.015	2.014101778	1+
Tritium	1	^{3}H	0 (12.32 y)	3.01602931	1/2+
Helium	2	^{4}He	100	4.00260325	0+
Lithium	3	^{7}Li	92.5	7.0116004	3/2-
Beryllium	4	^{9}Be	100	9.0121821	3/2-
Boron	5	^{11}B	80.1	11.009306	3/2-
Carbon	6	^{12}C	98.89	12.000000	0
Nitrogen	7	^{14}N	99.634	14.00307401	1+
Oxygen	8	^{16}O	99.762	15.99491462	0
Fluorine9		^{19}F	100	18.9984032	1/2+
Neon	10	^{20}Ne	90.48	10.99244018	0
Sodium	11	^{23}Na	100	229897697	3/2+
Magnesium	12	^{24}Mg	78.99	23.9850419	0
Aluminum	13	^{27}Al	100	26.9815384	5/2+
Silicon	14	^{28}Si	92.23	27.9769265	0
Phosphorus	15	^{31}P	100	30.9737615	1/2+
Iron	56	^{56}Fe	91.75	55.934942	0
Gold	79	^{197}Au	100	196.966552	3/2+
Lead	82	^{208}Pb	52.4	207.976636	0
Uranium	92	^{235}U	0.72	235.043923	7/2-
Uranium	92	^{238}U	99.274	238.050783	0

Table XII
Nuclear Masses

Atom	Atomic Number Electrons	Atomic Isotope	Atomic Mass Measured (amu)	Mass of Naked Protons & Electrons (amu)	Total Entron Mass (amu)	Entron Energy Mass (MeV)
Electron			0.000548	0.000584	0.0	0.0
Neutron	0	^0n	1.00866492	0.99883825	0.00982667	9.1585
Hydrogen	1	^1H	1.00782	0.99883825	0.00898175	8.3701
Deuterium	1	^2H	2.014101778	1.9976765	0.016425278	15.3084
Tritium	1	^3H	3.01602931	2.99651475	0.01951456	18.1876
Helium	2	^4He	4.00260325	3.9953530	0.00725025	6.7572
Lithium	3	^7Li	7.0116004	6.99186775	0.01973265	18.3514
Beryllium	4	^9Be	9.0121821	8.98954425	0.02263785	21.0985
Boron	5	^{11}B	11.009306	10.98722075	0.02208525	20.5529
Carbon	6	^{12}C	12.000000	11.986069	0.013931	12.9837
Nitrogen	7	^{14}N	14.00307401	13.9837355	0.01933851	18.0235
Oxygen	8	^{16}O	15.99491462	15.981412	0.01350262	12.5844
Fluorine	9	^{19}F	18.9984032	18.97792675	0.02047645	19.0840
Neon	10	^{20}Ne	19.9924402	19.976765	0.0156842	14.6177
Sodium	11	^{23}Na	22.9897697	22.97327975	0.01648995	15.3686
Magnesium	12	^{24}Mg	23.9850419	23.972118	0.0129239	12.0281
Aluminum	13	^{27}Al	26.9815384	26.96863275	0.01290565	12.0289
Silicon	14	^{28}Si	27.9769265	27.967471	0.0094555	8.8125
Phosphorus	15	^{31}P	30.9737615	30.96398575	0.00977575	9.1110
Sulfur	16	^{32}S	31.9720707	31.962824	0.0092467	8.6179
Iron	56	^{56}Fe	55.934942	55.934942	0.000000	0
Gold	197	^{197}Au	196.966552	96.7711353	0.1954167	182.13
Lead	208	^{208}Pb	207.976636	207.758356	0.21828	203.44
Uranium	235	^{235}U	235.043923	234.7269888	0.3169342	295.38
Uranium	238	^{238}U	238.050783	237.7235035	0.3272795	305.02

Electricity is flowing energetic conduction electrons (i.e. electrons with captured entrons). These conduction electrons can give up their entrons as photons in a light beam or as heat energy in an electric iron. These conduction electrons can also push naked electrons out of a coil of wire to produce an electric magnet or alternating magnetic fields in the coils of a motor or generator.

Magnetism is flowing naked electrons. Iron, cobalt and nickel atoms contain loosely attached naked electrons orbiting at their natural speed of 2.19×10^6 m/s. In a permanent magnet made of magnetic materials these naked electrons are coaxed to leave their atoms and travel through the magnetic material at their natural speed, to create the magnet, with the naked electrons exiting the magnetic material at its north pole and returning to the magnetic material at its south pole. Our Earth is a giant magnet. At 2.19×10^6 m/s it would take about three seconds for the naked electrons to make one cycle through our earth (South Pole to North Pole) and around the outside of our Earth back to the South Pole. The electrons may travel a little slower on the portion of their trip through the earth.

Alternating electric current is produced by using the naked electrons in a strong oscillating magnetic field to force conduction electrons to flow back and forth in a conductor.

CHAPTER XIV

ELECTRICITY AND MAGNITISM

What We Don't Know

In our modern societies almost everything we do involves electricity and magnetism. Many of our electronic gadgets and all of our motors and generators depend on the forces of both electricity and magnetism. Strange as it may seem, existing theories do not provide a clear description of either electricity or magnetism. If you do not believe it, GOOGLE either term. You will find that scientists do not know what an electron looks like or what it is made of. You will also discover that we know how to create magnetic fields, but no one knows what a magnetic field is. You may also discover that no one has a good explanation of voltage or charge. We all know that most of our home appliances operate with an alternating current at 120 volts AC or 240 volts AC and that our cameras, radios and cell phones operate from direct current batteries at lower DC voltages (such as 1½ volts, 3 volts and 9 volts). But what the heck is voltage anyway? The explanations that you will see talk about electric charge and the Coulomb force associated with the charge. But voltage represents energy, so what has charge got to do with energy? Don't get me wrong, scientist and engineers have learned how to deal with electricity and magnetism even though they do not understand it. They have developed mathematical formulas that make accurate predictions and these formulas work very well. We use them to design huge hydroelectric generators and microscopic integrated circuits. But scientists still do not know what an electron looks like and they certainly do not know what a magnetic field is nor do they know what an electric field is.

Magnetism

Magnetism according to the Ross Model is very simply nothing but naked (zero voltage) electrons circling through and around magnetic materials at the naked electrons' natural self-propelled speed of about 2.19 million meters per second (about 2.19×10^6 m/s). The naked electrons are traveling at the same speed as that of the orbital electrons of all atoms. The difference is the length of their paths. Our earth is a very big magnet produced by these naked electrons that loop through our earth, south to north, and around our earth, north to south, covering a distance of more than 6 million meters in about three seconds, making about 20 cycles per minute. (Electrons orbiting the nuclei of atoms (an orbit distance of about 5×10^{-10} m) do it in less than about 2.5×10^{-16} second, making about four thousand trillion cycles per second.) In permanent bar magnets these naked electrons flow freely through the atomic matrix of the magnet at their normal speed of about 2.19×10^6 m/s until the electrons reach the north pole of the magnetic material then the naked electrons continue at the same speed looping outside the magnet through a part of the wall of your refrigerator or through the space on the outside of your refrigerator and back to the south pole of the magnet.

There Are No Magnetic Monopoles

For more than a century scientists have been searching for a magnetic monopole which would be a north pole or a south pole not tied to an opposite pole. It is a wild goose chase since magnetic fields are circling zero voltage electrons. This means whenever these electrons leave magnetic material at a north pole, there has to be a south pole for the electron to return to.

Electric and Magnetic Fields

As suggested by the above discussions the magnetic fields are nothing but zero voltage electrons exiting a magnet at the north end and entering the magnet at the south end. The Ross Model proposes that electric fields are entrons drifting in a space between conductors held at different electric potentials. Or the electric field may be created by a very low current of energetic electrons flowing or tunnelling between the high potential and the low potential. The energy of energetic electrons is determined by the energies of their captured entrons.

Conduction Electrons

Photons, clectrons, positrons and protons have been described above based on the Ross Model. According to the Ross Model, the voltage (measured in volts) on an electrical conductor is determined by the energy (measured in electron-volts) of entrons captured by current carrying conduction electrons in the conductor. The conduction electrons of metals such as copper, silver, zinc and gold in their metallic form do not orbit the nuclei of the atoms of the metallic matrix, but they roam randomly through the metal matrix at their natural speed which is normally a little less than about 2.19×10^6 m/s. Low-voltage entrons are added to the metal as a part of the process of separating the metal from metal ores such as sulfides or oxides of the metals. Once the pure or semi-pure metal matrix is formed its conduction electrons can move freely through the metal matrix. These electrons collect entrons within the matrix and the entrons are shared with other conduction electrons so that very quickly all conduction electrons have entrons with energies distributed in a relatively narrow band of energy. In metals not a part of an electric circuit this narrow band of energy will correspond to the temperature of the metal. When a conductor is a part of a direct current electric circuit, current carrying electrons, for short periods, can be carrying entrons with much higher energies than all of the other electrons in the conductor. In alternating current circuits conduction electrons carrying entrons with energies ranging from zero to a maximum energy are forced to flow back and forth in the circuit. When an electron has captured two entrons the two entrons can change partners producing one high-energy entron and one low-energy entron. The high-energy entron may remain with the electron and the low-energy entron may be released back to the metal matrix as a heat energy entron or it may radiate away as a photon.

Naked electrons have a shape as shown in **FIG. 5** in **Chapter VII**. Its plus tronnie is traveling in a circle with a diameter of 0.934×10^{-18} m and its two minus tronnies circle through the center of that circle along circular paths that map out circular paths all of which have diameters of 0.934×10^{-18} m.

If a metal part is part of an electric circuit the energy of the current carrying conduction electrons is determined by the circuit. For example, in a metal conductor connected to the high-voltage terminal (the one not grounded) of a grounded 12-volt battery, current carrying conduction electrons have captured an entron with an energy of about 12 eV (relative to ground voltage) and a diameter of about 0.72×10^{-10} m, a little smaller than a typical atom. The ground voltage with an earth ground will be based on the temperature of the ground. At standard temperatures, 300 K (23 C), according to Wien's Law (see Table V in Chapter I), as interpreted by the Ross Model, entrons corresponding to this temperature have energies mostly in the range of about 0.13 volts, so the entrons of the conduction electrons at ground voltage will be entrons with an energy of about 0.13 eV. The diameter of a 0.13 eV entron would be about 6.7×10^{-10} meters, about 6.7 times the size of a typical atom.

Metals such as copper, zinc, silver and gold are very rarely found in nature as pure metals. This is because their last one or two electrons are far distant from the positively charged nucleus and being repelled by the orbital electrons circling below. These electrons are normally traveling too fast (about 2.19×10^{6} m/s) to remain tied to the nucleus of the atom and are lost leaving the atom as a positively charged ion which tends to combine with an atom such as sulphur or oxygen both of which has two empty spaces in their outer orbit and are happy to accept two additional zero voltage electrons traveling at 2.19×10^{6} m/s since the other six electrons in the outer orbit are already traveling at that speed.

The high-voltage terminal of a 2.3 volt battery would provide energetic electrons with entrons having diameters of about 3.77×10^{-10} m. Entrons in these low-voltage ranges are about 190 million times larger than electrons that have captured them. When the 2.3 eV energetic electrons flow through a resistor (such as a very thin tungsten filament in an incandescent lamp), all or almost all of the entrons are lost from the electrons. The very much smaller naked electrons (stripped of their entrons) have no trouble passing through the filament and flowing on to the low-voltage terminal of the battery. The filament temperature is increased to about a five thousand degrees Kelvin. At 5,000 degrees Kelvin, entrons have a broad range of energy between about 1.5 eV and about 2.8 eV and peaking at about 2.1 eV and the filament avoids even higher temperatures by releasing some of these entrons. See **Table V** in **Chapter V**. So a small percentage of these entrons (about **5 percent**) are released as visible light photons. Most of the entrons are released or are conducted away as thermal entrons or infrared or ultraviolet photons not within the visible spectrum. If the lamp is a 10-watt lamp, the electric energy provided to the lamp will be 10 joules per second, which equals about 6.24×10^{19} eV/s. Visible light photons have energies in the range of about 2.3 eV. So even with an efficiency of only 5 percent the 10 watt incandescent lamp will produce about 3×10^{18} photons per second. That is about 3 million trillion photons per second. A 100 watt lamp would produce about 30 million trillion photons per second!

When entrons are lost in a light emitting diode (an LED) they escape as photons with energies corresponding closely with the supply voltage of the LED. With a proper design these LED's release most of their entrons within the visible spectrum and can have efficiencies in the range of about **70 percent**! LEDs are constructed with an N-doped region constructed to permit the transit of energetic electrons and a P-doped region constructed to permit the transit of zero voltage electrons with an intrinsic translucent region sandwiched in between. Energetic electrons from the high voltage terminal of the battery easily pass through the N-doped region into the intrinsic translucent region, but they can pass through the P-doped region only by losing their entrons which exit the intrinsic translucent region as visible light photons. The now naked electrons pass through the P-doped region and on to the low voltage terminal of the battery. The naked electrons then pick up an entron from a chemical reaction in the battery and exit the high-voltage terminal as an energetic electron to complete the circuit.

When batteries are connected in series, entrons representing the battery voltage add to the energy of the electrons at the low voltage terminal of each battery, so three 12-volt batteries in series will produce electrons with entron energies of about 36 eV. Let us consider three 12-volt batteries (A, B and C) connected in series. At the low-voltage (grounded) terminal of 12-volt battery A, each conduction electron will have absorbed an entron with a diameter of about 6.7×10^{-10} m and an energy of about 0.13 eV (corresponding to the ground voltage which in turn depends on the temperature of the grounded terminal). At the high-voltage terminal of battery A all conduction electrons will have captured an entron with an energy of about 12 eV and a diameter of about 0.72 $\times 10^{-10}$ m. At the low-voltage terminal of a 12 volt battery B, all current carrying conduction electrons will have captured an entron with about the same energy and diameter as the electrons at the high-voltage terminal of battery A (i.e. 12.13 volts and 0.72×10^{-10} m). The 12.13 eV electrons at the high-voltage terminal of battery A do not pass through battery B. Instead zero voltage electrons in the copper wire connecting battery A to battery B enter battery B and capture a 12 eV entron from battery B. The 12.13 eV entrons are left at the high-voltage terminal of battery A and the low-voltage terminal of battery B. Similarly these high-energy electrons at the high-voltage terminal of battery B do not pass through battery C. Instead zero voltage electrons in the copper wire connecting battery B to battery C enter battery C and capture a 12 eV entron from battery C. The 24.13 eV entrons are left at the high-voltage terminal of battery B and the low voltage terminal of battery C. The electrons at the high-voltage terminal of battery C will be at a potential of 12 volts relative to the high-voltage terminal of battery B but at a potential of 24 volts relatively to the high voltage terminal of battery A and at a potential of 36 volts relative to ground.

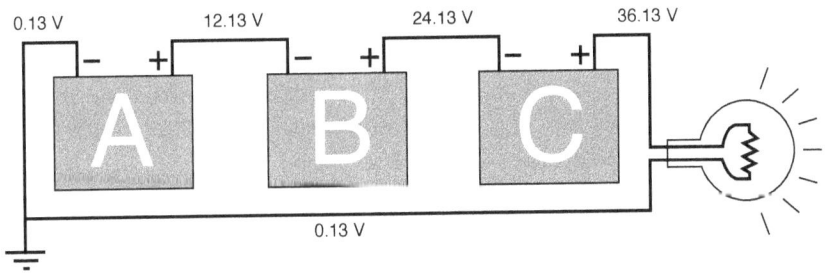

Electric Current

At normal temperatures all electrons associated with the atoms of metal conductors (other than conduction electrons) are in their ground states and are orbiting around the nuclei of a metal atom. These electrons typically have no captured entrons and in the Ross Model they are called "naked electrons" as explained above. They are circling the nucleus at a velocity of about 2.19×10^6 m/s. Conduction electrons in a metallic atomic matrix typically are not attached to any particular atom and are traveling freely in the metal matrix of atoms. The conduction electrons typically have captured at least one entron and would be traveling at a velocity different from 2.19×10^6 m/s. The entrons captured by conduction electrons define the electron's energy and the voltage of the circuit. The tronnies of low-energy entrons (less energetic than 13.6 eV) loop through electrons and the Coulomb forces from the looping tronnies provide a backward force on the electrons to slow down the electrons to reduce the electron velocity and kinetic energy. A 13.6 eV entron reduces the electron velocity to a velocity relatively close to zero. Very high-energy entrons give the electron a large velocity that can be any velocity up to 1.414 times the speed of light in a direction opposite the direction of its naked electron.

Electrons that have captured entrons with energies above 13.6 eV but less than about 27.2 eV may be able to continue circling the nuclei of non-metal atoms with velocities lower than about 2.19×10^6 m/s, and if so they would be in excited states and their orbits would be larger than the orbit radius of a naked electron. They drop from their excited states to a lower excited state or ground state by releasing the entrons (often as photons) and if they release all of their entrons to return to the ground state they speed back up to 2.19×10^6 m/s. I have a lot more to say about excited states in my discussion of the **Ross Atom** in **Chapter XXVI** which is a take-off of the **Bohr Atom** in which I examine the spectral lines of the hydrogen atom.

Electron Flow

Although electric current can be the flow of any type of charge, in most siturations electric current represents a flow of electrons, and we normally measure that electric current in amperes (amps). If you look back at **Table V** in **Chapter I** you will see that an ampere is equal flow of one coulomb of charge per second. But a Coulomb of charge is equivalent to the combined charge of 6.2415×10^{18} electrons! That is 6.2415 billion billion electrons. So 0.04 amp of electric current which is the electric current that would power a small LED would represent a flow of about 0.25 billion billion electrons per second! If the little LED were provided with a 2.4 volt forward bias it would produce red light photons with a peak wavelength of about 0.660×10^{-6} m (equivalent to entron energies of about 1.88 eV (see **Table 5** in **Chapter V**). Assuming an efficiency of about 50 percent would mean this little LED would be emiting about 1.6×10^{18} photons per second. That would be about 1.6 billion billion photons every second!

The Copper Atom

An electrical conductor such as a copper wire and at least two sources of electrons at different electric potentials are required to provide an electric current. According to the Ross Model the nucleus of each copper atom has a net positive charge of 29e (where e is the charge of a single electron and the magnitude of the charge is about 1.602×10^{-19} coulombs). There are two stable isotopes of copper, copper-63 (69.17 %) with 63 protons and 34 electrons in its nucleus) and copper 65 (30.83%) with 65 protons and 36 electrons in its nucleus). Both isotopes have 29 additional electrons associated with its nucleus but are outside the nucleus. Only 28 of these electrons are in fixed orbits around the nucleus: 2 at a first orbit pattern, 8 at a second orbit pattern and 18 at a third orbit pattern. All of these electrons are naked electrons self-propelled at their natural speed of 2.19×10^6 m/s, and each are tightly synchronized with each other to avoid getting any closer together than necessary. (Electrons hate each other and repel each other with enormous forces if they are close together.) At these speeds and with their mass of about 9.1 X 10^{-31} kg, each electron has the same average momentum. Their centripetal force is dependent on their orbit radius. That centripetal force on the average over time must exactly match the Coulomb force applicable between each electron and the nucleus and the other electrons especially those in a lower orbit rings. Each electron wants to orbit as closely as it can to the nucleus, but since they are all traveling at the same extremely high speed (almost one percent of the speed of light); space is limited in the orbit patterns. Once the lower patterns are full, additional electrons are forced to try to form additional rings. However, after each ring becomes full, a single electron attempting to form another ring may find that its natural speed of 2.19 X 10^6 m/s gives it a centripetal force greater than the net Coulomb force resulting from the nucleus and the electrons orbiting below it. So for copper atoms the 29th electron cannot orbit the copper nucleus at its natural speed. This is why copper in nature is almost always in an ionic form combined with another atom such as oxygen or sulphur which has happily agreed to take the extra electron to become a negative ion and to form an oxide or sulphide with copper.

However there is a way that copper and other conductive materials can exist as a pure material. That is to slow down the extra electron (the 29th electron in the case of copper) so that its centripetal force is equal to or less than the net Coulomb force of the copper nucleus and the electrons orbiting at the lower rings. The extra electron is slowed down by capturing an entron. The entron, remember, is two opposite trommies circling with a diameter corresponding to the energy of the entron. For example, a 2 eV entron has a diameter of about 4.3 X 10^{-10} m, about the size of a typical atom.

Let us look at **Equation 5** from **Chapter IX**.

$$v = \sqrt{\frac{2(E_N - E_e)}{m_N + m_e}}$$

The mass of an entron with energy in the range of 2 eV entron (at about 3.56 X 10^{-36} kg) is very small relative to the zero energy electron mass (at about 9.1 X 10^{-31} kg) so the equation reduces to:

$$v = \sqrt{\frac{2(E_N - E_e)}{m_N}}$$

where E_N is 13.6 eV and E_e is the energy of the entron and m_N is the electron mass.

A 2-volt potential applied to an electrode terminal can provide a copper atom with a 2 eV entron. If you plug in these numbers (converting electron volts to joules, you will see that the entron has sufficient energy to slow the electron down from 2.19×10^6 m/s to about 2.02×10^6 m/s. A 4-volt potential would provide a 4 eV entron that would slow the electron down to about 1.8×10^6 m/s and a 10-volt potential would slow the electron down to 0.79×10^6 m/s. With the entron providing a breaking action on the electron it is less eager to escape metal copper atom, and the result is that the 29th electron may be able to orbit a copper nucleus in an orbit outside the other 28 electrons. Once copper atoms are assembled in a matrix of only copper atoms not connected to an electric circuit the 29 electron typically travels freely through the matrix and typically does not leave the metal matrix. Even if the conduction electron loses its entron and reverts to its natural speed of 2.19×10^6 m/s, it can remain in the copper matrix traveling randomly at that speed. It is welcome to remain in the matrix since each conduction electron provides a negative charge that neutralizes one of the copper ions in the matrix each of which has a charge of plus one. Not only that, this conduction electron can exchange its entron for entrons of any energy (such as an energy of 0.13 eV corresponding to thermal energy at 25° C) up to thousands or millions of electron volts. As shown in **FIG. 8** in **Chapter IX**, entron energies in excess of 13.6 eV will tend to increase rather than decrease the electrons speed.

Electric Generator

As explained earlier in this chapter, the Ross Model proposes that magnetic fields are zero energy electrons that flow out of the North poles of magnet and pass through space and flow back into the South poles of either the same magnet or other magnets in series. These electrons according to the Ross model are all traveling at their natural speed of about 2.19×10^6 meter per second outside and maybe a little slower while inside the magnet.

Conduction electrons in all conductors (except possibly conductors at extremely low temperatures) are energetic electrons. All of the current carrying conduction electrons in a conductor will have captured an entron corresponding to the electrical potential of the conductor. That potential will be at a ground potential corresponding to the temperature of the conductor or if the conductor is part of an electric circuit, the current carrying conduction electrons will be at a potential as determined by the circuit.

If a conductor is placed in a magnetic field, the zero voltage electrons in the magnetic field will pass through the conductor. The zero-energy electrons in the magnetic field will disturb the balance of plus and minus charges in the conductor causing some of the energetic electrons and some of the zero energy electrons in the conductor to move out of

114

the way of the zero energy electrons in the magnetic field. If the magnetic field is not changing and the conductor is not moving relative to the magnetic field, equilibrium will quickly develop in the conductor. If in a single conductor conduction electrons in a portion of the conductor are at a different potential compared to conduction electrons in another portion of the conductor, the conduction electrons will quickly (at speeds of millions of meters per second) mix with each other sharing their captured entrons until all conduction electrons are at approximately the same potential. That is very quickly they will all have captured entrons with approximately the same energy. We see that sharing of entron energy as electric current flowing from high potential regions to low potential regions. However, if the field is changing or the conductor is moving in the magnetic field, a changing current of conduction electrons will be set up in the conductor. Also if a current of energetic electrons is flowing in a conductor, the flow of conduction electrons will disrupt the balance between plus and minus charges in the conductor. As a current of conduction electrons flows in a copper wire, zero energy electrons will be forced out of the wire. These zero-energy electrons tend to circle the wire creating a magnetic field around the wire. By arranging the wire in coils powerful magnetic fields can be created with a flow through the coil of conduction electrons. A rotating magnet will produce a rotating magnetic field and the rotating magnetic field will produce an oscillating current of conduction electrons in a conductor positioned in the oscillating field. Also an oscillating current of high-energy conduction electrons can be applied to produce oscillating magnetic fields in the center of coils of conductor material.

Albert Einstein was not a supporter of the uncertainty principle, which proposes that we cannot know the product of position and velocity to accuracies better than h/2π (i.e. 1.055 X 10^{-34} kgm^2/s). The Ross Model sees no need for uncertainty principle. The Ross Model has no explanation for the uncertainty principle, but it is interesting that product of the momentum of the electron in the Bohr atom (the hydrogen-1 atom) and the radius of the electron's path is approximately equal to h/2π.

CHAPTER XV

THE UNCERTAINTY PRINCIPLE

Uncertainty

The prior art idea that uncertainty is a basic element of physics is wrong according to the Ross Model. There may be things going on that we have no way of measuring precisely, but physics is completely precise. Coulomb's law and its derivations apply to infinitely small dimensions and at all speeds. The fundamental particles in our Universe are point particles so their positions (wherever they are) should be infinitely precise. Coulomb's law is infinitely precise, at least as far as we know. Time is absolute. There is no limit to how many times you can divide a second. So our Universe should be infinitely precise. Currently, our instruments are not infinitely precise, but they are getting better and better.

Heisenberg's uncertainty principle states that a minimum exists for the product of the uncertainties in position and momentum that is equal to or greater $h/2\pi$, where h is Planck's constant, $h = 6.626 \times 10^{-34}$ kgm^2/s and π = about 3.1416. So this uncertainty value is:

$$\text{Uncertainty} = (6.6260755 \times 10^{-34} \text{ } kgm^2/s)/(2)(3.1415928) = 1.05312 \times 10^{-34} \text{ } kgm^2/s$$

Just for the fun of it let us see if we can examine a naked electron using Ross Model values for the naked electron. Its natural velocity is 2.190877×10^6 m/s and its mass is 9.1094×10^{-31} kg. So we know its momentum (mass times velocity) $p = 19.9529 \times 10^{-25}$ kg-m/s. We also know that when the electron is a ground state electron in orbit around a proton in a hydrogen atom its radius (the Bohr radius) $r = 5.29177 \times 10^{-11}$ m. So we know if we multiply the radius by the momentum of the electron we get the uncertainty value (i.e. $h/2\pi = 1.0207 \times 10^{-34}$ kgm^2/s); i.e.

$$rp = (5.29177 \times 10^{-11} \text{ m})(19.929 \times 10^{-25} \text{ kg-m/s}) = 1.0207 \times 10^{-34} \text{ } kgm^2/s,$$

which is very close to 1.05312×10^{-34} kgm^2/s

I have no idea what the meaning of this is. I just find it very interesting. Also we may examine the neutrino photon in terms of measuring its momentum and its position. My estimated diameter of the neutrino photon circle is about 0.853×10^{-15} m so its radius is 0.4265×10^{-15} m. Its mass is about 1.65×10^{-27} kg and its velocity is about 3×10^8 m/s so its momentum (mv) is about 4.9638×10^{-19} kg m/s. The product of its diameter and its momentum is about 2.12×10^{-34} kg m^2/s. Half of this value is very close to Heisenberg's uncertainty value. Also, I have no idea what this means.

If the basic charged particle were not a point particle, it would blow itself apart in order not to violate Coulomb's Law.

All composite particles with a net charge other than zero and with a size other than zero must be comprised of at least three particles with charge; and in the three-particle combination, the charge of at least one of the particles must be opposite the other particles.

The simplest composite particles with a net charge are the electron and the positron.

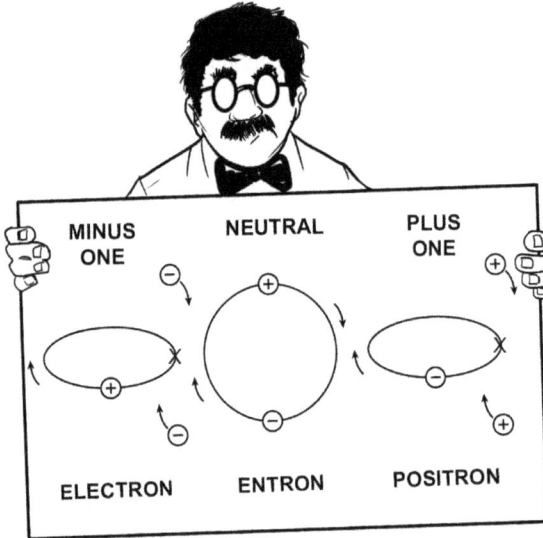

MINUS ONE   NEUTRAL   PLUS ONE

ELECTRON   ENTRON   POSITRON

CHAPTER XVI

TRONNIES ARE REQUIRED BY COULOMB'S LAW

Coulomb's Law

Coulomb's Law says that the force between two stationary charges is proportional to the product of the charges and inversely proportional to the square of the distance between them. Ross' Law says that the integrated force between two circling tronnies is inversely proportional to the distance between them. In either case at distances infinitely close to zero, the Coulomb force approaches infinity (i.e. a portion of the charge would be extremely close to a second portion each with the same charge). A consequence of both of these laws is that all elementary particles with a single charge must be point charges with no volume or any other dimension. Another consequence is that all composite charged particles must be comprised of point charges. Otherwise, the elemental particles or the composite particles would blow themselves apart with infinite or near infinite force. It is obvious that a point cannot have mass. Infinite density makes no sense. These point charges are tronnies. They are the source of the Coulomb force and the building blocks of our Universe. An important question is: "Could an elementary charged point particle travel at speeds less than the speed of light?" The Ross Model answer is "No". Once a point particle is going away from a past position of itself faster than the speed of light its own Coulomb force waves from its own past (especially its immediate past) will keep it going at at least the speed of light. Therefore, we can assert that every elementary charged particle in our Universe must be a point particle doomed to always travel at speeds equal to or greater than the speed of light.

As we will see in **Chapter XXV**, I now propose that tronnies must travel in precise circles at speeds of $\pi c/2$ (about 1.57c) with respect to at least one frame of reference, and each tronnie is the precise point focus of its own Coulomb force waves coming diametrically across its circle from its recent past. (I am not sure whether they <u>always</u> travel in precise circles or <u>almost always</u> travel in precise circles.) As the Coulomb waves from the opposite side of the circle pass through the tronnie's focus point, the waves spread out in all directions

The Ross Model supports conservation of energy except in the course of Big Bang explosions. The question is, "If there is conservation of mass-energy, how could our Universe have gotten so big?" During the course of the Big Bang explosion that was the death of our predecessor universe and the birth of our Universe, the mass-energy of our Universe may have increased such as by a factor of 2 as compared to our predecessor universe. The Ross Model does support the concept that mass-energy cannot be destroyed. But definitely mass-energy can be created. Otherwise our Universe could not have become so big!

CHAPTER XVII

CONSERVATION OF MASS-ENERGY

Entrons Have Mass

The Ross Model supports the concept of conservation of mass-energy with a couple of reservations. The Ross Model recognition, that entrons have mass and that each photon is comprised of one entron, means that photons have mass. Thus, when a body cools by radiating photons, the mass of the body is decreased by the mass of the radiated photons. When a uranium atom fissions in a nuclear reactor the mass of its fission products is less than the mass of the original uranium atom by the mass of the gamma rays released in the fission process. Following the Big Bang naked protons were created by the combination (in each case) of an electron, two positrons and a neutrino entron. Then in the subsequent weeks, months and years the naked protons collected gamma ray entrons from gamma ray photons: (1) to help cool down our Universe and (2) to slow down to become hydrogen nuclei. The masses of the protons were increased by an amount equal to the masses of the collected gamma ray entrons. These gamma ray entrons in the hydrogen nucleus are currently being released in fusion processes by stars to provide light and energy to their surrounding planets. The hydrogen atoms in our Sun are being fused into helium atoms. Helium nuclei do not need as much gamma ray entron energy to slow down enough to become helium atoms. The excess entrons are released as photons to provide life giving energy and mass to our earth.

Tree leaves collect entrons produced in our sun and transferred to our earth as photons. These entrons are combined with carbon and other atoms and molecules to produce wood which is burned in our fire places releasing some of the same entrons collected by the leaves of the trees.

Protons are destroyed in Black Holes at the center of each galaxy and the massive neutrino entron in each proton is released to help provide the gravity of the galaxy. This is the ultimate example of conversion of mass into radiation energy and a great example of the conservation of mass-energy since the masses of the neutrino entron, the proton's electron, two positrons and any leftover gamma ray entrons all together is going to equal the mass of the destroyed proton.

All of the above supports to concept of conservation of mass-energy. That is, you lose some energy and you gain an equivalent amount of mass and you lose some mass and you gain an equivalent amount of energy, all according to $E = mc^2$.

121

How Did Our Universe Get So Big?

There is one thing however that does bother me about conservation of energy-mass. The Ross Model proposes that our Universe was created in the Big Bang explosion that was the death of our predecessor universe. And that there has been a series of universes preceding our predecessor universe going all the way back to some series of events that preceded the first universe.

The Ross Model proposes that the very first universe was a conglomeration of mass-energy and that this first universe was relatively small, at least as compared to our Universe, perhaps the size of our Milky Way Galaxy (i.e. about 2×10^{42} kilograms). And each successive universe is much larger in terms of mass-energy than its predecessor, such as by a factor of two. If so, after a series of about 47 universes, our Universe could be enormously more massive (i.e. in the range of 1.46×10^{53} kilograms).

If there were a law of conservation of mass-energy that applies even through Big Bang events, it would mean that all of our predecessor universes must have contained the same mass-energy as our Universe. I find this difficult to explain. I would rather believe that in Big Bang explosions there is an enormous creation of new mass-energy. This could result from the compression of entrons. The mass of each entron is inversely proportional to its diameter. If the diameters of entrons are compressed, additional mass-energy is created. The Ross Model proposes that each Big Bang explosion has been an explosion of a Monster Black Hole that has with its enormous gravity pulled together all or mostly all of the galaxies in its universe. The compression of billions of galaxies in a single Monster Black Hole could result in the compression of entrons. This could explain substantial increases in the energy-mass of each successive universe.

If you will refer to **Table V** in **Chapter V** you will see that at a temperature of 2.16 trillion degrees Kelvin all entrons and photons are neutrino entrons and neutrino photons, each having a mass-energy approximately equal to the mass-energy of a proton. We can imagine that in the collapse of the Monster Black Hole that preceded the Big Bang that was the birth of our Universe, all or many of the entrons at the central portion of the Black Hole were squeezed down to the size of neutrino entrons.

The Ross Model proposes that entire universes can be assembled from nothing but tronnies. This model also proposes that the tronnies produce Coulomb force waves and tronnies are the focus of Coulomb force waves. So all we need to build a universe are either tronnies or Coulomb force waves. So which came first the chicken or the egg? Which came first the tronnies or the Coulomb waves? I do not have the answer to where the first tronnies or the first Coulomb force waves came from. But it seems clear to me that they came from somewhere. It is also clear to me that tronnies and Coulomb force waves can combine under certain circumstances to produce additional tronnies. Once tronnies are created they produce Coulomb force waves forever that expand forever. These Coulomb force waves under special circumstances should be able to come to focus to produce other tronnies. Therefore, it is easy at least for me to believe that each successive universe can be substantially more massive than its predecessor universe violating the law of conservation of mass-energy.

At the very beginning, long before the first universe, before there was anything, there was nothing not even tronnies or Coulomb force waves. But now we have a universe of somewhere between 100 billion galaxies and 400 billion galaxies each with an average of about 100 billion stars, most of which have planets with moons. This does not suggest a **conservation** of energy-mass; this does suggest the **creation** of energy-mass.

Each photon is a combination of a particle (its entron) and waves (Coulomb waves produced by the tronnies of its entron). Thus a photon is both a particle and waves. Photons in photon beams ride in Coulomb grids produced by the tronnies in the photon beams and the tronnies in the cosmic background radiation and other radiation. Photons can be reflected or refracted by other Coulomb grids, both stationary grids and grids moving at the speed of light. In two-slit experiments a single entron passes through only one slit, but its Coulomb waves pass through both slits. Photons react with matter via reflection, refraction, the photo-electric effect, Compton scattering or electron-positron pair production.

CHAPTER XVIII

PHOTON INTERACTIONS

Particles or Waves

Scientists do not know whether a photon is a wave or a particle or both. The Ross Model says a photon is both a wave and a particle. The entron portion of the photon is a particle in the same sense that an electron is a particle. The entron represents the mass of the photon. The entron can exist in many forms other than as a component of a photon. It can be captured by an electron or a proton to become part of the particle giving the electron or the proton electrical energy and additional mass. It can be trapped in matter in the form of a heat quantum, later to be released as part of a photon.

But of the two tronnies in each entron of each photon in a beam of light and all the other entrons in the beam of light (with their charges) produce Coulomb force waves that travel out from the tronnies' positions to produce a Coulomb wave structure in which the entrons in the beam of light ride. The entron of a single photon can create a complex wave traveling at the speed of light in which the photon can travel at the speed of light through a laboratory or through our Universe. The photon is the combination of the entron particle and the Coulomb wave the entron produces and in which the entron rides. So the photon is both a particle and a wave!

Reflection and Refraction

In **Chapter XXIII**, I propose that photons travel through Coulomb grids and that these Coulomb grids are produced by all of the tronnies in the space through which photons are traveling. So let us imagine that any particular Coulomb grid can constitute a reference frame from which we can view the photon. The looping path of the entron viewed from the Coulomb reference frame through which the photon is traveling at the speed of light has the speed of the entron varying from plus 3c to minus c while the photon travels forward at the speed of c (see **FIG. 4** in **Chapter V**). The entron travels backwards through a significant portion of each wavelength. At surfaces such as the surface of a lake, a mirror, a window or a tree leaf, Coulomb fields flow out from the surface at the speed of light perpendicular to the surface. Therefore it is easy for some of the entrons to change directions at surfaces. Entrons floating in a beam of light intersecting these surfaces at an incident angle float in the Coulomb field flowing out from the surface as well as the Coulomb field of its own incoming beam so they often reflect from the surface at a reflection angle equal to the incident angle. Entrons illuminating a material of higher index of refraction that are not reflected or absorbed near the surface of a material flow with its wave into the material at a slower speed bending toward the normal to the surface in a wave-like pattern.

A "Single Photon - Two Slit" Explanation

One of the fundamental mysteries of physics is the single photon, two slit experiment. In this experiment single photons (coming one at a time) illuminating two slits spaced close together in a first screen produce a diffraction pattern on a second screen. This mystery is described in many physics books. For example, Fundamentals of Physics, Halliday, Resnick and Walker, Sixth Ed., John Wiley & Sons, Inc. at pages 962-964. The question is how can single photons produce diffraction patterns? How could a photon passing through one of the two slits even be aware that there is a second slit? The best explanation the prior art can give is that the light is produced as a photon and is absorbed as a photon but "travels between the source and the detector as a probability wave". (The Ross Model does not support probability waves.) The Ross Model provides a much simpler solution. The entron portion of each photon passes through only one of the two slits, but its Coulomb wave passes through both slits and the entron portion combines with its own Coulomb force wave that passes through both slits to help direct the path of the entron portion of the photon on the opposite side of the screen. So the entrons collect in a pattern on the screen that is partially determined by the Coulomb waves that pass through both slits.

Polarization

The two tronnies of each entron travel in circles in a single plane. Each photon is comprised of a single entron traveling in the plane of its entron spin and the direction of the photon is in that same plane. In most beams of light the spin plane of the entron and its photon is random about the photon direction (the photon axis). Light beams become partially polarized upon reflection from smooth surfaces. I have tried many explanations of polarization but none seem really simple. I have also tried to understand polarization under other theories of light. None really make much sense to me. Light can be polarized by a number of techniques other than reflection and the polarization of a light beam can be changed with wave plates and the polarization of the beam can be rotated with magnetic fields. I believe my model can explain polarization but it is probably going to take someone much more expert in optics than I who is a believer in tronnies and entrons to provide that explanation.

Photon Absorption

When photons react with matter, they may be reflected or transmitted without loss of energy. They may also be absorbed to create an excited atom or molecule. And they may be absorbed in the matter as heat energy. High-energy x-rays and gamma rays are attenuated via the photo-electric effect, Compton scattering or pair production. It is the entron that is the energy/mass of the photon. When the photon is trapped in matter for a short time or for a very long time, it ceases to exist as a photon but its entron does not cease to exist. It can however transfer part of its energy to other entrons or the tronnies of entrons can change partners with the tronnies of another entron. For example, more than one entron can be captured by the same electron. When this happens, the tronnies of the one of entrons is brought closer to the tronnies of a separate entron than it is to its partner. The result can be a higher energy entron and a lower energy entron. In a microwave oven entrons of a single relatively low frequency (long wavelength and low energy) become absorbed in what is cooking and the result is a hot product with a wide range of entron frequencies, wavelengths and energies some of which are much higher energy entrons. Some of these high-energy entrons are absorbed in the material being cooked changing it molecular structure. Other entrons are radiated away as photons or conducted away as entrons as the product cools. As explained in **Chapter VI**, three entrons (a neutrino entron at 928 MeV, a 1.02 MeV gamma ray entron and a 1.12 KeV entron) can combine to produce an electron and positron pair in the process called "pair production". Most entrons trapped in matter as heat energy radiate from the matter in the form of photons. In a reaction described as the photoelectric effect, the entron portion of the photon is absorbed by an electron giving the electron all of the entron energy. This may cause the electron to be ejected from the matter in which it is located. In Compton scattering the entron is absorbed by an electron, but an entron previously a part of the electron is ejected and we monitor the ejected entron as a lower energy photon.

Heat and temperature are nothing but entrons. The diameters of the entrons decrease as the temperature increases. At a temperature of freezing and melting water, the diameter of the entron is about 7.42×10^{-9} m (about the size of a water molecule). At the boiling point of water the entron diameter is about 5.47×10^{-9} m.

The entrons in our bodies that are keeping us at a temperature of 98.6 °F have a diameter equal to about 6.53×10^{-9} m, about the size of a small molecule.

CHAPTER XIX

HEAT AND TEMPERATURE

Heat and Temperature Provided by Entrons

Heat is nothing more than entrons (circling tronnie pairs) temporarily trapped in matter. Temperature is a measure of that heat. We warm up when our bodies absorb the infrared, visible and ultraviolet light entrons of photons radiated from a camp fire or the sun, and we cool down when our body radiates photons with wavelengths in the infrared and millimeter wave range.

A 400-watt microwave oven produces about 250 trillion very low-energy entrons per second. These low-energy entrons of the microwave radiation each with an energy of about 1.02×10^{-5} eV corresponding to an extremely low temperature of less than 1.0 K warm our TV dinners to about 373 K (100 °C). Each of these microwave entrons has a diameter in the range of about 8.5×10^{-5} m and the microwave photon has a wavelength in the range of about 0.12 m. We typically must wait a few minutes for some of the 373 K entrons to radiate away (each one carrying away energy of about 0.16 eV) so that the dinners have cooled enough for us to eat them, i.e. to about 310 K (37 °C and 98.6 F). Entrons that radiate out of the TV dinner at 310 K (98.6° F) have energies that peak at about 0.133 eV (see **Table V**), a diameter of about 6.51×10^{-9} m and their photons are infrared photons with a wavelength of about 9.32×10^{-6} m.

So the question is how can these very low-energy entrons, 1.02×10^{-5} eV, produce the infrared 0.133 eV entrons that radiate from the TV dinners with energies about 100,000 times greater and with diameters 100,000 times smaller? The answer has to be that the tronnies of the microwave entrons must trade partners so that some of them produce combinations of much higher energy entrons. At the rate of 250 trillion entrons per second, in two minutes 30 thousand, trillion entrons will be dumped into this small oven. The tronnies of these entrons are forced close together forming entrons with diameters much smaller than the diameters of the microwave entrons.

Table V in **Chapter V** contains a comparison of the entron diameters d' to the peak temperature T (in degrees Kelvin) of a substance. The relationship is:

$$d' = 2.025 \times 10^{-6} \text{ mK/T}$$

So in a substance at a temperature of 0° C (273 K) the diameter of contained entrons is in the range of about 7.42×10^{-9} m and at 100° C (373 K) the diameters are about 5.43 X

10^{-9} m. Therefore liquid water contains entrons with diameters in the range of about 5.43 X 10^{-9} m to about 7.42 X 10^{-9} m.

Our bodies are continuously absorbing entrons (mostly in the form of photons from our environment) and we are continuously radiating entrons away from our bodies mostly in the form of infrared photons. The entrons in our bodies that are keeping us at a temperature of 98.6 °F have a diameter equal to about 6.51 X 10^{-9} m, about the size of a small molecule. Our lives depend on our 6.53 X 10^{-9} m diameter entrons. If we deviate more than a few degrees from 98.6 °F for any substantial period our bodies cease to function.

Prior art models include phonons which are referred to as quasi-particles to describe quanta of heat energy. Most scientists do not claim the phonons are real; instead they describe them a "wave functions". The Ross Model has no need for phonons or these wave functions. Entrons are the basic quanta of heat energy. Entrons are real, just as real as a photon. In fact a photon is an entron traveling at the speed of light as explained above. When photons are absorbed in your skin the photons cease to exist but their entrons continue to exist within your body as heat energy or the photon could be absorbed by a skin molecule of your body to change the structure of the cell to help give you a nice tan. Or they could be utilized by your body in the very many processes your body uses to keep you alive. It is the entrons within your body that keep you at the correct body temperature and some of them are continuously being radiated away to maintain your body at about 98.6 °F.

Helix Nebula

Image credit: NASA/JPL-Caltech

The Ross Model predicts that Black Holes operate at temperatures much greater than the temperatures of the hottest stars. Protons are destroyed in Black Holes at the center of each galaxy, releasing a neutrino entron with each proton destroyed. The neutrino entrons escape as neutrino photons to provide the gravity of the galaxy.

A small portion of the neutrino photons are temporarily trapped in stars, planets and moons to provide these bodies their gravity.

The destruction of an earth size planet each day by the Black Hole at the center of our Milky Way Galaxy would produce a neutrino photon flux, at our earth, of about 300,000 photons/m^2-second. This means that about 150,000 neutrino photons from the Black Hole in the center of our galaxy may be passing through each of our bodies each second as a part of its process of holding our galaxy together. Additional neutrino entrons from our Sun are passing through our bodies as part of the process by which our Sun is holding our earth and the other planets in orbit around our Sun. And more neutrino photons are passing through our bodies each second as a part of the process by which our Earth is keeping us from floating away.

Raise your hand if you think you can feel any of these neutrino photons passing through your body.

CHAPTER XX

BLACK HOLES AND GRAVITY

Black Holes

Each galaxy in our Universe contains at least one Black Hole at or near its center which provides sufficient gravity to hold together the entire galaxy of many billions of stars and their planets and the moons of the planets. Matter within each galaxy, including the stars, planets and moons, is continually being consumed by the Black Hole. Once consumed we never see the matter again. Scientists do not have a good explanation as to what happens to the consumed matter or a good description of gravity. And they do not have a good description of the processes going on inside Black Holes. The Ross Model provides the answers.

The Ross Model proposes that temperature inside Black Holes is much higher than temperatures inside the cores of stars. The matter consumed by the Black Hole is broken down, by this tremendous heat of the Black Hole, into atoms and the atoms are broken down into electrons, protons and entrons. Protons are then destroyed by their combination with anti-protons. Each destroyed proton releases an electron and two positrons and one neutrino entron and additional lower energy entrons. Each of the destroyed anti-protons produces two electrons and one positron and one neutrino entron and additional lower energy entrons. The released lower energy entrons add heat energy of the Black Hole to help maintain its tremendously high temperature. The Ross Model proposes that there is within each Black Hole an enormous flux of neutrino entrons produced from destruction of the protons. Combinations of neutrino entrons, positrons and electrons produce a large population of anti-protons at a controlled rate that depends on the population of positrons, electrons and neutrino entrons within the Black Hole. Each anti-proton soon after production combines with a proton and both are annihilated releasing two neutrino entrons, some of which participate in the production of more anti-protons and some of which escape the Black Hole as neutrino photons to provide the gravity of the galaxy.

The neutrino photons travel out from the Black Hole at the speed of light. Coulomb force waves spread out from each neutrino photon at the speed of light. (Remember a neutrino photon is a single neutrino entron traveling in a circle at a speed of 2c within the neutrino photon which is traveling at a speed of c.) Neutrino entrons are so small (with a diameter of about 0.934×10^{-18} m), that almost all of their photons pass easily through all objects in their path including objects as small as atoms (with sizes of about 10^{-10}) and even the nuclei of atoms (with diameters of about 1,000 times larger at about 10^{-15} m) or as large as planets and stars. Therefore, objects in the path of the neutrino photons do not feel the

Coulomb forces from the neutrino photons until the neutrino photons have passed by or through the objects. Thus, there is no force on the objects in the direction of travel of the neutrino photons. Sidewise forces produced by Coulomb waves from the neutrino photons cancel so the only net force felt by the objects, through which the neutrino photons are passing, is a force pushing the objects back toward the source of the neutrino photons. This is described in more detail in the next section entitled "Gravity". The flux of neutrino photons leaving the Black Hole, near the surface of the Black Hole, is so intensive that normally nothing other than the neutrino photons can leave the Black Hole. All other photons in the electromagnetic spectrum, including visible light, infrared light, ultraviolet light radio waves, x-rays and gamma rays, are all forced back into the Black Hole by the Coulomb waves from the neutrino photons. The Ross Model proposes that all neutrino photons have the exact same energy and wavelength and they do not share their energy with each other or other photons, except in the process of pair production and the generation of protons and anti-protons. Neutrino photons and neutrino entrons are about 100 times smaller and more massive than the next highest energy entrons. The Ross Model proposes that photons other than neutrino photons slow down or speed up when they travel into a Coulomb field that is moving relative to the source of the photons so as to always travel at the speed of light relative to the Coulomb field through which the photon is traveling. Neutrino photons however always travel at the speed of light relative to their source. This distinction is very important for the understanding of gravity in accordance with the Ross Model.

Gravity

Neutrino entrons released from the Black Hole at the center of each galaxy produce the gravity of the galaxy. Here is how:

Atoms, including their nuclei and the protons in the nuclei, according to the Ross Model, are comprised of nothing but electrons, positrons and entrons. The size of electrons and positrons is about 2×10^{-18} m, which is about a thousand times smaller than a proton and one hundred million times smaller than a typical atom. As explained above the neutrino entron (which is the mass and energy of each neutrino photon) has a diameter of about 0.934×10^{-18} m (about half the size as the electron and the positron) and it spins with the same frequency as the electron. Electrons and positrons are the only things that can capture a neutrino entron. Since neutrino entrons and the electrons and the positrons are so small, the probability of a capture in a finite volume of matter is extremely small, but luckily not zero. So the neutrino photon easily passes through objects like stars, planets and moons, molecules, atoms and even protons, but a small percentage of the neutrino entrons are temporally captured by electrons and positrons in objects such as the stars, planets and moons. Most of these captured neutrino entrons are later released in random directions as neutrino photons; making these stars, planets and moons sources of neutrino photons and giving the stars, planets, and moons their gravity. Coulomb force effects from the tronnies in each neutrino photon produce tiny forces on the charges in the objects through which the neutrino photons pass, pushing the objects back toward the source of the neutrino photons (i.e. toward Black Holes, which are the primary source of neutrino photons, or the stars, planets and moons which are the secondary sources). Electrons in hydrogen atoms in interstellar space capture or scatter a portion of the neutrino photons to significantly reduce the neutrino photon flux from each galaxy that reaches far distant galaxies.

This Ross Model explanation of gravity is going to be challenged by the scientific community which has been lead to believe that gravity results from the curvature of space produced by massive objects such as our sun, our earth and the moon. The Ross Model proposes that space is empty space. It is nothing. It can't be curved. The Ross Model of gravity is much simpler. Massive objects capture neutrino entrons produced in Black Holes at the center of each galaxy and later release them in random directions producing the gravity of the massive objects. The larger the mass the more neutrino entrons it captures. The more it captures, the more it later releases, so the gravity of each object is proportional to its mass. Neutrino entrons from the Black holes do good by producing the neutrino photons that are holding the stars, planets and moons in the galaxy in their assigned orbits around the Black Holes. However, the Black Holes, in order to do its job, do bad by consuming portions of their galaxies to produce the neutrino photons.

The reader is encouraged to examine the path taken by the entron in each photon (**FIG. 4** in **Chapter V**). You will note that once during each cycle of the photon the entron travels at speeds that vary from minus c to plus 3c. The neutrino photon is one half the size of an electron. It has no trouble passing through stars, planets and moons but it can occasionally be captured by an electron and later released. At some point someone needs to make a computer model that will simulate neutrino photons passing through objects such as our sun, our earth and our moon. The model should examine the forces acting at distances in the range of 10^{-18} m and specifically examine the Coulomb forces of each of the two tronnies in the neutrino entron. The model should demonstrate how the Coulomb forces from the neutrino entrons are able to apply forces to the atomic and sub-atomic particles in these objects in order to direct them back toward the source of the neutrino photons. Many other computer models are needed to confirm the many predictions of the Ross Model. For 101 predictions see **Chapter XXIX**.

How Many Neutrino Photons Are Produced by Black Holes?
Black holes regularly consume an entire solar system its sun, its planets and moons. I am going to guess that the Black Hole in the center of our galaxy consumes matter at an average rate of about one earth size planet each day. Let us as an example estimate the

number of neutrino photons that are released from a Black Hole with an average consumption per earth day of a single planet the size of our Earth. A planet the size of our earth would have a mass of about 5.98×10^{24} kg. Since almost all of that mass of the planet is provided by protons and since each proton has a mass of about 1.67×10^{-27} kg, there must be about 3.6×10^{51} protons in a planet the size of the earth. Since each proton is comprised of one neutrino entron, the planet would carry into the Black Hole about 3.6×10^{51} neutrino entrons which sooner or later would be converted to neutrino photons. (Remember the number 3.6×10^{51} neutrino photons.) So let us try to get an idea of the extent of the neutrino photon flux at our solar system as a consequence of the daily destruction of an earth size planet by the Black Hole at the center of our Milky Way Galaxy. To begin with our earth is about 2.2×10^{20} meters from the location of the Black Hole at the center of our galaxy.

The surface area A of a sphere with a radius of 2.2×10^{20} m is determined by the formula: $A = 4\pi r^2$, where r is the radius of the sphere, so the surface area of a sphere at the solar system's position in the Milky Way Galaxy is about 6.1×10^{41} m^2. We assume that all of the neutrino photons released by the destruction of a planet the size of the earth would be distributed evenly over the surface area of a sphere with a radius equal to the distance between the solar system and the center of our galaxy. Now we want to know the number of neutrino photons per square meter at the earth's average position in the galaxy. We get that number by dividing 3.6×10^{51} neutrino photons by 6.1×10^{41} m^2. The result is 0.59×10^{10} neutrino photons/m^2 or about 5.9 billion neutrino photons per square meter. So if we assume the Black Hole consumes matter at the rate of **one earth-size planet each day** and converts its protons to neutrino photons, our earth should see a neutrino photon flux of about 5.9 billion neutrino photons per square meter per day or if we divide this number by the number of seconds (86,400) in a day; we have a neutrino flux of about **68,000 photons/m^2-second**. This may provide a rough estimate of the neutrino photon flux we are actually experiencing that is holding our solar system in its circular path around the Black Hole at the center of our Milky Way Galaxy. Since our bodies have a cross section of about 0.5 square meter (with these assumptions) we, ourselves are being penetrated each second by about **34,000 neutrino photons** each second from the Black Hole in the center of our galaxy! (And keep in mind this does not count the neutrino entrons passing through our bodies secondarily from our sun, our moon and our earth. Although it is not pleasant to think about being riddled by so many neutrino photon "bullets" each second of our lives, we need to keep in mind that were it not for these tiny neutrino photon bullets; we, our earth and our sun would not be maintained in our respective positions in our Universe. In fact it is difficult to imagine what our Universe would be like without these tiny gravity producing neutrino photon bullets.

Dark Energy

While we are on the subject of neutrino photons and gravity, I might bring up the subject of dark energy. Existing theories concerning gravity and the current expansion of our Universe have suggested that there may be a lot of dark energy in our Universe that we have been unable to detect and a lot of matter that we cannot see. The problem that these scientists have is that they are not aware of the neutrino photons of the Ross Model. Existing theories do include the old fashion "neutrino". But old fashion neutrino is not supposed to have any mass or if it does its mass is supposed to be very small. For the Ross Model lack of dark energy and invisible mass is not a problem. You can see in the preceding paragraph the neutrino photons represent a tremendous quantity of dark energy and dark matter. Each neutrino photon has a mass almost equal to the mass of a proton. And our galaxy is loaded with them. Also many escape from our galaxy and other galaxies to load up intergalactic space with invisible neutrino photons providing dark energy and dark matter.

The example described in the preceding section demonstrates how Black Holes are continuously converting real visible matter (such as an earth-size planet) into dark matter such as 3.6×10^{51} neutrino photons per day. The example assumes an average conversion of one earth-size planet per day for a typical galaxy. Remember we have about 100 to 400 billion galaxies in our Universe and the Black Holes in the galaxies have been making these conversions continuously every day for many billions of years. The result is a heck of a lot of dark energy and dark matter.

The Ross Model provides another source of matter that is not dark, but is not currently considered in the calculation of the total mass of our Universe. This is photons other than neutrino photons. The Ross Model proposes that all photons have mass, even radio wave photons and that the mass of each photon is proportional to the energy of the photons. All of this is very important in connection with the future of our Universe and questions of whether it will continue to expand forever, expand for some period then stop expanding or expand for some period and the contract into a small volume and explode in a future Big Bang. We will deal with these issues in the next chapter and **Chapter XXVI**.

Why Are Black Holes Black?

We estimated a neutrino flux of roughly $68,000/m^2$ neutrino photons at a distance of 2.2×10^{20} meters from the Black Hole at the center of the Milky Way Galaxy. This would mean that at distances much closer to the Black Hole the neutrino flux would be so large that the neutrino photons would be almost touching, providing a virtual wall of speed of light neutrino photons providing a reverse wall of Coulomb forces pushing everything else back into the Black Hole.

Anti-gravity results from low energy photon pressure on far-away galaxies and maybe from reflection of neutrino photons from the shell of our Universe.

We have prepared estimates indicating that photon pressure on our Milky Way Galaxy from faraway galaxies could be in the range of roughly 1.28×10^{31} newtons. The radiation pressure per square centimeter is extremely small but the cross section of our galaxy is extremely large. We have calculated a very small resulting acceleration of only 0.514×10^{-10} meters per second squared. But during a period of one billion years, this could result in an increase in the velocity of our galaxy of 1.619×10^{6} meters per second. This represents an expansion of more than $1\frac{1}{2}$ million meters per second.

CHAPTER XXI

ANTI-GRAVITY

Photon Pressure

According to my Ross Model there are two sources of anti-gravity. These sources are low-energy photons and reverse direction neutrino photons. Low-energy photons such as visible light photons pass through intergalactic space basically unimpeded, much more efficiently than gravity producing neutrino photons. Therefore, low-energy photon pressure from stars of one galaxy is sufficient to provide a significant repulsive accelerating force on far distant galaxies. This force per square meter is small but galaxies are big and the force is continuous (always accelerating every second) for billions of years. For galaxies that are close to each other, gravity producing neutrino photons trump low-energy photons, but for far-away galaxies low-energy photons trump neutrino photons. Thus, close-by galaxies are all accelerating toward each other due to the influence of neutrino photons which pass through the stars, planets and moons of the close by galaxies, and far-away galaxies are expanding away from each other due to pressure from lower energy photons which are absorbed by or reflect from the stars, planets, moons, dust and other matter of the accelerated galaxies. So the things responsible for the attraction and repulsion of galaxies are photons, nothing but photons, low energy photons pushing far-away galaxies apart and neutrino photons pulling close-by galaxies together!

Rough Estimate of Anti-Gravity from Photon Pressure

We can make a rough estimate of this anti-gravity repulsive force. For example, we can see dimly some far-away galaxies with our naked eyes. So we know that the approximately 1 square centimeter apertures of our eyes must be collecting some amount of visible light photons each second from the far-away galaxy. Let us assume that the number is 100 visible light photons per second. There are 10 thousand square centimetres in a square meter. Therefore we estimate the flux of visible light photons from this particular far away galaxy is roughly 1×10^6 (i.e. $100 \times 10,000$) photons per second per square meter. Each visible light photon carries a quantum of energy of about 2.29 eV. One eV is equal to 1.6×10^{-19} watt-seconds. So the energy transmitted from distance galaxies to our galaxy by visible light photons may be in the range of about 3.66×10^{-13} W/m^2. This radiation power per square meter is also called flux density. So the flux density produced by visible light radiation from a far-away galaxy in the range of about:

Visible light flux density = about 3.66×10^{-13} W/m^2 = about 3.66×10^{-13} Nm/sm^2

since 1 watt = 1 Nm/s.

We need to relate this flux density to force or pressure. It is known that if radiation is absorbed the pressure is the flux density divided by the speed of light. If reflected the pressure is doubled. So if we assume that all of the visible light radiation is absorbed, our estimate of radiation pressure on our galaxy from visible light from a far-away galaxy could be in the range of about:

$$\text{Radiation pressure} = (3.66 \times 10^{-13} \text{ Nm/sm}^2/(3 \times 10^8 \text{ m/s}) = 1.22 \times 10^{-21} \text{ N/m}^2$$

This is an extremely small pressure, but our galaxy is very big. The diameter of our Milky Way Galaxy is about 1×10^{21} meters so the cross sectional area of our galaxy is about $(0.5 \times 10^{21} \text{ m})^2 \times (3.1416) = 0.7 \times 10^{42} \text{ m}^2$. Let us assume that about 5 per cent of the photons that attempt to traverse our galaxy are absorbed. (We are assuming for simplicity zero reflection.) If so the effective cross section of our galaxy for absorption of visible light photons from a single far away galaxy is in the range of about:

$$\text{Cross section} = (0.05) \times (0.7 \times 10^{42} \text{ m}^2) = 0.035 \times 10^{42} \text{ m}^2 = 3.5 \times 10^{40} \text{ m}^2$$

So to get the force on our galaxy produced by visible light from a far-away galaxy, we multiply the effective cross section of our galaxy to visible light by the radiation pressure. So our estimate of the force is:

$$\text{Visible light radiation force} = (1.22 \times 10^{-21} \text{ N/m}^2)(3.5 \times 10^{40} \text{ m}^2)$$
$$= 4.27 \times 10^{19} \text{ N}$$

This is about 42.7 billion billion newtons produced on our galaxy from visible light from a single far-away galaxy.

For galaxies located near the center of our Universe radiation pressure from all far-away galaxies would cancel and not contribute to the velocity of the galaxy. So let's assume that our galaxy is located near an edge of our observable universe in the direction that puts it closer to the shell of our Universe than to the center of our Universe. In this case the radiation pressure on the galaxy from all far away galaxies should accelerate our galaxy in a direction away from the center of our Universe toward its shell. The question is how much?

There are more than 100 billion galaxies in our Universe. Let's assume that 75 billion of those galaxies push our galaxy toward the shell of our Universe and 25 billlion push it toward the center of our Universe. So we have a net of 50 billion galaxies pushing toward the shell. So how much is the acceleration? First let's multiply our pressure from one galaxy by 50 billion. Then let's multiply the result by 3 based on an assumption that electromagnetic radiation other than visible light radiation will contribute to the pressure on our galaxy. These assumptions increase our estimate of the radiation pressure as follows:

$$\text{Radiation force} = (4.27 \text{ X } 10^{19} \text{ N}) \text{ X } (50 \text{ X } 10^{9}) \text{ X } (3) = 2.13 \text{ X } 10^{30} \text{ N}$$

Now let's divide this number by 2 to account for the fact that most of the radiation is not directed along the acceleration direction. So our estimate of the acceleration force in the direction of acceleration is:

$$\text{Radiation force in acceleration direction} = 1.065 \text{ X } 10^{30} \text{ N}$$

This is a tremendous force. But is it enough to accelerate a galaxy to any significant extent? Isaac Newton taught us that $f = ma$. The mass of our galaxy is estimated at about $1.25 \text{ X } 10^{12}$ solar masses and one solar mass is about $1.99 \text{ X } 10^{30}$ kg. So the mass of our galaxy is about $2.49 \text{ X } 10^{42}$ kg. So we can estimate the acceleration that results from photon pressure between far-away galaxies as:

$$a = f/m$$
$$\text{Acceleration} = (1.065 \text{ X } 10^{31} \text{ N}) / (2.49 \text{ X } 10^{42} \text{ kg}$$
$$= (1,065 \text{ X } 10^{30} \text{ kg m/s}^2) / (2.49 \text{ X } 10^{41} \text{ kg}) = 2.65 \text{ X } 10^{-11} \text{ m/s}^2$$

This is a very tiny value, but the acceleration is approximately continuous for billions of years, and remember there are more than 31.5 million seconds in a year and more than 31.5 million-billion seconds in a billion years. So let's estimate the change in velocity of this galaxy if it is accelerating at this rate for one billion years. We know that the change in velocity is the product of acceleration and time. So:

$$\text{Velocity after one billion years} = (2.65 \text{ X } 10^{-11} \text{ m/s}^2) \text{ X } (31.5 \text{ X } 10^{15} \text{ s})$$
$$= 8.35 \text{ X } 10^{5} \text{ m/s} = \text{about 835 km/s}$$

This is a little more than one half of one per cent of the speed of light. Astronomers tell us that the velocity of our galaxy relative to the cosmic background radiation is 552 +/- 6 km/s.

With the gross assumptions that I made in attempting to calculate the velocity of our galaxy, I did not expect to come very close to estimating the actual speed of our galaxy, but I did expect to be in the ball park, and I think I am! To me anti-gravity is primarily due to photon pressure between far-away galaxies.

Reverse Direction Neutrino Photons

According to the Ross Model neutrino photons are indestructible once created they live forever. There is one exception as explained in **Chapter VI**, a neutrino photon is consumed in a pair production process where the neutrino photon and two other photons combine to produce and electron and a positron. However, as also explained in **Chapter VI** the neutrino photon is quickly recreated when the positron combines with an electron to reproduce the neutrino photon and two gamma ray photons. Neutrino photons can also be captured by an electron for a short period after which they are released in random directions. It is hard to tell whether these stray neutrino photons are producing gravity or anti-gravity. However many neutrino photons make it to the shell of our Universe (see the next chapter, **Chapter XXII**). Some of the neutrino photons combine with an electron two positrons to produce protons at the outer regions of our Universe as explained in **Chapter XXIII**. But about half of the neutrino photons that are released at the shell head back toward the galaxies of our Universe. As these neutrino photons pass through galaxies they produce forces directed toward the shell of our Universe. This is a second source of anti-gravity.

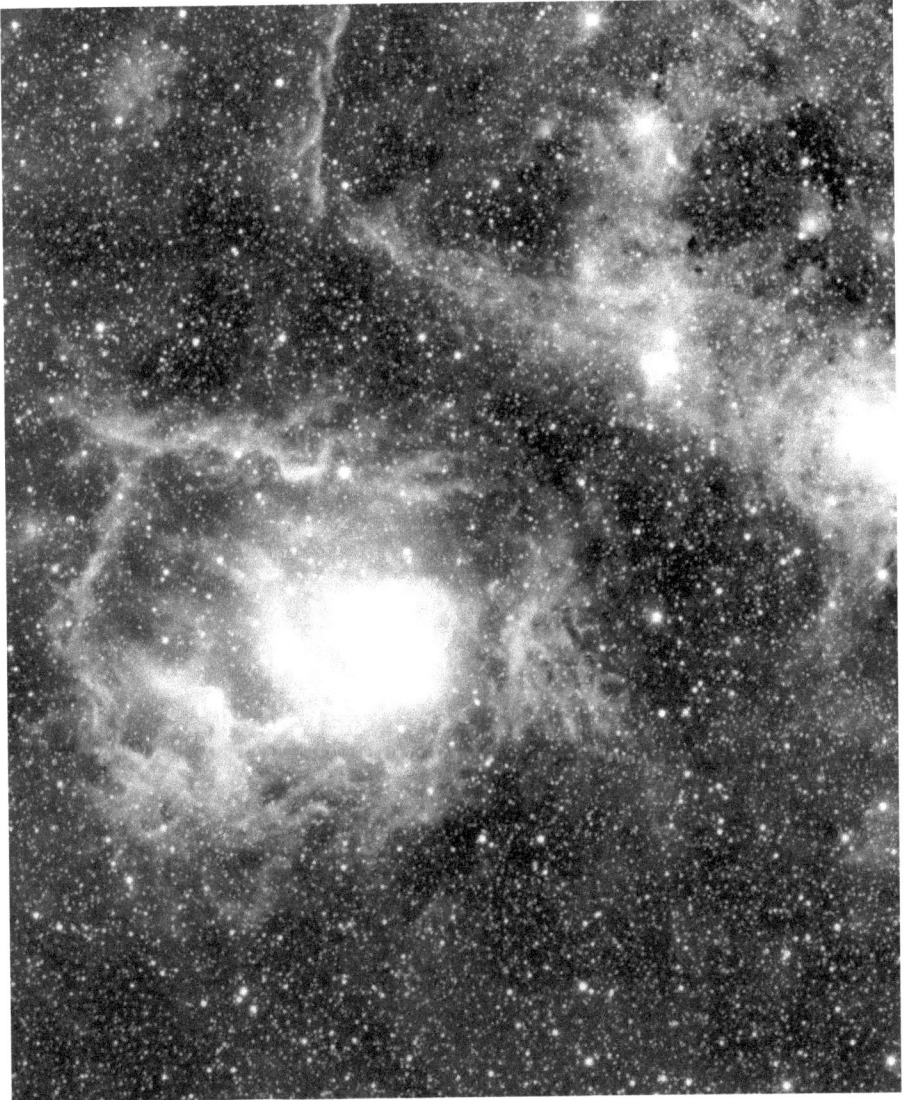

Lagoon Nebula

Image credit: NASA/JPL-Caltech

Our Universe is surrounded by a cold plasma shell of mostly naked electrons and naked positrons left over from our predecessor universe. These naked electrons and positrons are all traveling at 2.19×10^6 m/s, but they do not normally combine because they are traveling too fast. The cold plasma shell adsorbs high-energy photons and reflects low-energy photons to produce the cosmic background radiation.

The high-energy radiation combines to produce additional electron-positron pairs at the inside surface of the shell and additional protons in the interior of the shell to ultimately create new galaxies at the outer edges of our Universe. Some of the neutrino photons are temporally captured and released in the shell and directed back toward the center of our galaxy to assist in the production of the anti-gravity that is producing the expansion of our Universe.

CHAPTER XXII

COLD PLASMA SHELL OF OUR UNIVERSE

Electron-Positron Plasma

The Ross Model describes a cold plasma shell) that surrounds and contains our Universe. The model proposes that this shell is many light years thick and is comprised mostly of naked electrons and naked positrons that were left over from our predecessor universe or were created shortly after the Big Bang that gave birth to our Universe. Almost all of them are traveling at 2.19×10^6 m/s and spaced widely apart so collisions and annihilations are relatively rare.

Cosmic Background Radiation

This cold plasma shell has reflected low-energy photons produced by the stars of the galaxies of our Universe like an integrating sphere since the formation of our Universe and it has absorbed or reflected neutrino photons and high energy photons. (An interesting device is the integrating sphere. It is a sphere-shaped shell with a highly reflective inside surface for producing a photon flux that is uniform everywhere inside the shell.) Because of the shell of our Universe, photons from every star in our Universe do not escape from our Universe. Over billions of years the shell has reflected radiation directed away from the galaxies of our Universe back into our Universe. These multiple reflections tend to create a uniform background of cosmic radiation. We call this background of cosmic radiation the "cosmic background radiation". This radiation of reflected low-energy photons is almost completely uniform in every direction. This cold plasma shell is also the birth place of new galaxies. The electrons in the cold plasma shell absorb the entrons of neutrino photons (neutrino entrons) reaching the shell to produce very high-energy electrons relatively stationary and circling with a diameter of 0.85×10^{-15} m. The high-energy electron circling with a diameter of 0.85×10^{-15} makes a much better target for positrons as compared to the naked electrons (a thousand times smaller) traveling at almost one percent of the speed of light, and each energetic electron may capture two positrons to produce a new proton which collects gamma rays to become a hydrogen nuclei which in turn can collect an electron to become hydrogen atom providing material for the production of new galaxies at the currently growing boundary of our Universe.

Anti-Gravity Produced in the Shell

Many of the electrons in the shell that capture neutrino entrons will decay before they capture two positrons. Each decay releases the neutrino entron in random directions. A neutrino photon is also produced with each electron-positron annihilation in the shell. These neutrino entrons could be directed into our Universe to contribute to the anti-gravity of our Universe discussed in the previous chapter.

The cold plasma shell of our Universe as explained above is comprised primarily of naked electrons and naked positrons all traveling at their natural speed of 2.19×10^6 m/s. The outer portion of the shell is probably almost entirely zero energy naked electrons and naked positrons traveling at 2.19×10^6 m/s. The inner layer of the shell is probably at the temperature of our Universe which would give the electrons and positrons some electric energy corresponding to their captured entrons. The inner portion of the shell also probably includes some gamma ray photons and entrons and some neutrino photons and entrons, along with lower energy entrons. Some protons, naked and energetic are probably also present in the lower layers of the shell. Pair production and electron positron annihilation are probably taking place in the lower layers of the shell. In between the inner layers of the shell and the outer layers of the shell probably includes a large graded layer with temperatures ranging from the temperature of our Universe to absolute zero at the outer most layer. The Ross model proposes that the shell is very thick. In fact the shell may have a thickness that is larger than the diameter of our Universe which it is containing. The reader is warned at this point that the Ross Model is going to propose that the shell is going to play an important role in explaining the inflation period of our Universe associated with Big Bang and the beginning of our Universe.

Sword of Orion

Image credit: NASA/JPL-Caltech/Univ. of Toledo

Our Universe is 100 percent empty space. This is because everything in our Universe is made from tronnies that occupy no space. However our Universe is completely filled by a luminiferous ether made up of Coulomb grids consisting of Coulomb waves traveling at the speed of light in all directions.

Light beams travel through the Coulomb grids at the speed of light. In a typical relatively low-energy light beam the density of the photons may be in the range of about 2×10^{11} photons per cubic meter. That is 0.2 trillion photons per cubic meter. But each photon is comprised of only one entron, which is two tronnies, each of which are point particles. So light beams occupy no space. On the other hand, the Coulomb force waves from the tronnies in the beam completely fill the space through which the light beams pass.

We have also shown in this chapter that an eight cubic centimeter cube of "solid" copper metal is 100 per cent empty space.

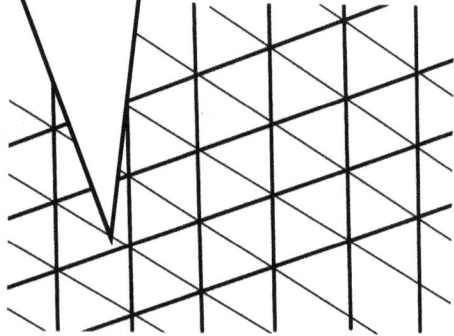

CHAPTER XXIII

COULOMB GRIDS

The Luminiferous Ether

As I have emphasized many times our Universe is comprised of nothing but tronnies and things made from tronnies. Each tronnie is a point charge, either plus or minus. And each of these charges travel in a circle at about 1.57 times the speed of light and carry the Coulomb force which comes to a focus in the tronnie and then travels out from the tronnie as continuously expanding waves at the speed of light applying the Coulomb force (attractive or repulsive) to themselves and other tronnies. Composite particles include the three tronnies of electrons and positrons and the two tronnies making up the entron in each photon. Protons, atoms and molecules all are comprised of nothing but tronnies and things made from tronnies. In portions of our Universe where a lot of mass is concentrated the aggregation of all of these tronnies create Coulomb force waves that create thick Coulomb grids. In other portions of our Universe the Coulomb grids are relatively very sparse. But there is no place in our Universe that is not completely filled with Coulomb force waves which define Coulomb grids. These Coulomb grids are all comprised of speed-of-light Coulomb waves mostly traveling in random directions. It is these Coulomb grids that permit and support the passage of photons. It is these Coulomb grids that are in many ways equivalent to the "luminiferous ether" that was proposed by scientists to support the passage of light before the Michelson & Morley experiment seemed to disprove the existence of the luminiferous ether. This historic luminiferous ether was supposed to be constant throughout the universe and galaxies, stars and planets were supposed to be moving through the luminiferous ether. The Ross Model Coulomb grids differ in that all significant bodies moving through our Universe carry with them their own Coulomb grids. As will be explained in the next chapter, **Chapter XXIV**, photons need a Coulomb grid for support because each of the two tronnies (that make the entron that is the energy of each photon) is a point focus of Coulomb force waves and depends on the Coulomb grid to support their charges of plus e or minus e.

Ross Model Relativity

We can imagine simple and complicated Coulomb grids. For example, imagine a cubic meter in interstellar space far away from any galaxy. The only tronnies in this space would be those contained in the few atoms of hydrogen that float around in interstellar space (estimated to be about one hydrogen atom per cubic meter) plus the tronnies in photons of a variety of energies from the far away galaxies and randomly directed photons of the cosmic background radiation. The combined Coulomb force waves at any time from the tronnies in this cubic meter of space would set up a Coulomb grid that on the average is approximately stationary relative to the center of our Universe and its shell; however, the grid is made up of Coulomb waves all moving at the speed of light in a variety of directions which are approximately random. Let's refer to this grid as our "universal Coulomb grid" and for the time being, let's assume that it truly is universal, i.e. approximately the same everywhere in our Universe that is far away from any galaxy. We should understand however that this grid is expanding, but let's avoid that issue for now. Next let's imagine a cubic meter of space located within our Milky Way Galaxy but far away from any star system. This cubic meter would be similar to the cubic meter we just considered except there would be substantially more photons passing through and probably significantly more hydrogen and other atoms floating around, so the Coulomb grid in this cubic meter would be denser. A very significant difference however would be that most of the photons would be coming from the 100 billion stars within the galaxy all of which are moving through the galaxy at high speeds around the center of the galaxy and the galaxy itself is moving through the universal Coulomb grid at a high speed. A similar situation would apply to other galaxies in our Universe. So the Coulomb grids within almost all galaxies are all moving at high speeds with respect to the universal Coulomb grid. And since most galaxies are moving at high speeds with respect to each other, their Coulomb grids are moving at high speeds with respect to each other. Also our sun and our solar system and all of the atoms and photons associated with our solar system are moving through our Milky Way Galaxy at a high speed relative to the Coulomb grid of our galaxy. Therefore, a Coulomb grid is defined by our solar system and this Coulomb grid is moving through space at a high velocity along with our solar system. And lastly our earth is moving through the Coulomb grids of the solar system, the Milky Way galaxy and our Universe at a high speed relative to those grids. All of the atoms of our earth and its atmosphere and the photons produced by or reflected from our earth and its components define "an earth Coulomb grid" that is moving through space at the same speed that our earth moves through space. You may ask, "Why is all of this important?" It is important because according to the Ross Model, visible light travels through Coulomb grids at the speed of light (c = about 3×10^8 m/s) relative to the Coulomb grid it is passing through. Therefore, if a beam of light passing through interstellar space at a speed of 1.0 c illuminates a galaxy moving at a speed of 0.1c opposite the beam of light, the beam of light will slow down to 0.9 c while it passes through the galaxy. Anyone in the galaxy measuring the speed of the beam will measure its speed a 1.0 c. On the other hand if the galaxy is moving through space at a speed of 0.1c in the direction of the light beam the light beam will speed up to 1.1c while it passes through the galaxy but the people in the galaxy measuring its speed will also measure its speed as 1.0 c. (The same analysis applies to star systems and planets within each galaxy.) The same analysis applies to measurements of the speed of light on earth. Earth carries its own Coulomb grid so any light passing through the earth's atmosphere will slow down or speed up so that its speed as it passes through is the speed of light (i.e. 1.0 c or about 3 X 10^8 m/s).

Light Beams Create Speed of Light Coulomb Grids

According to the Ross Model photons are entrons traveling in a circle at a speed of 2c and forward at the speed of c all as shown in **FIGS. 3** and **4** in **Chapter V**. As photons in a beam of light pass through space at the speed of light, the tronnies of each of their entrons are producing Coulomb waves that create a Coulomb grid moving at the speed of light through the Coulomb grid the light beam is passing through. How can this be? Let's figure. How many tronnies are there in a typical beam of light? Let's take a beam of visible light created by a 100-watt incandescent light with a reflector large enough to create a one square meter light beam. Let's assume that the light system is 20 percent efficient so the combined energy of all of the photons is 20 watts. (A watt is equivalent to 1.0 joule per second, 1.0 J/s). The energy of visible light photons is about 2 eV/photon or 3.2 X 10^{-19} J/photon. This means that this beam of light must have a flux of about 6.25 X 10^{19} photons/s-m^2.

The beam is moving at the speed of light (3 X 10^8 m/s) so in the beam the density of photons must be about:

$$photons/m^2 = \frac{6.25 X 10^{19} \, photons/s - m^2}{3 X 10^8 \, m/s} = 2.08 X 10^{11} \, photons/m^3$$

Each photon is comprised of two charges, each charge equal to 1.6 X 10^{-19} coulomb, one plus and one minus. Thus, a typical one square meter light beam from a 100 watt lamp produces a density of about **400 billion charges per cubic meter** moving through space at the speed of light. These 400 billion charges per cubic meter create their own Coulomb grid that is moving at the speed of light through space. It is this Coulomb grid that functions as a river carrying the circling entrons through space as a beam of light. If this beam were a perfect laser beam, the beam could travel through space for billions of years; however, light beams, even laser beams, sooner or later tend to spread out and become less dense.

Our Universe
Empty Space Filled Completely with Coulomb Waves

The Ross Model proposes that everything in our Universe is made from tronnies or things made from tronnies and that tronnies are point particles with no volume. It also proposes that every tronnie produces Coulomb waves that expand out form the tronnies at the speed of light, each wave expanding forever. If these proposals are correct, then our Universe must be **100 percent empty space** which is completely filled with Coulomb waves. I know this result sounds ridiculous, because among other things it means that you, the reader, and I are 100 percent empty space along with our earth and everything else.

But let us take a close sub-atomic look at a small piece of our Universe. Let us examine an 8 cubic centimeter ($8 \text{ cm}^3 = 8 \times 10^{-6} \text{ m}^3$) block of copper metal. If we conclude that it is 100 percent empty then maybe our entire Universe (including us) is all empty space. This cube of copper (with 2 centimeter edges) has a mass of about 63.5 grams which is about 1 gram-atom of copper. That means the cube contains about 6×10^{23} copper atoms. That is 600 billion trillion copper atoms. In copper metal the atoms are packed close together, each atom occupying a volume of about:

$$V_{atom} = 8 \times 10^{-6} \text{m}^3 / 6 \times 10^{23} \text{ atoms} = 13 \times 10^{-30} \text{ m}^3.$$

The cube root of this volume gives the spacing of the atoms in the copper cube which is about 2.35×10^{-10} m.

But we know each atom is comprised of a small massive nucleus with orbiting electrons. So there must be a relatively lot of empty space within each copper atom. Let us see how much. First let us determine the volume of the nucleus. The size of the nucleus of atoms can be estimated using the formula:

$$r = r_0 A^{1/3}$$

where r is the radius, r_0 is 1.2×10^{-15} m and A is the atom's mass number, 63.5 for copper. So the radius of the nucleus of the copper atom is about:

$$r = 4.78 \times 10^{-15} \text{ m}$$

The nucleus is approximately round so the volume can be estimated using the formula:

$$V = (4/3)\pi r^3 = (4/3)(3.1416)(4.78 \times 10^{-15} \text{ m})^3 = 20 \times 10^{-45} \text{ m}^3.$$

The volume of each of the 29 electrons is very small. Existing theories do not agree on the size of electrons, but they do agree that they must be very small since electrons flow right through the matrix of thin copper wire with no trouble. The Ross Model proposes that electrons have a radius of about 2×10^{-18} m, so if the electron were a round solid with this radius it would have a volume of about:

$$V = 8 \times 10^{-54} \text{ m}^3.$$

Since each copper atom has 29 electrons associated with it, the electrons would have a total volume of about:

$$V = 232 \times 10^{-54} \text{ m}^3$$

which is completely insignificant compared to the volume of the nucleus. So the total volume occupied by the copper atoms (their nuclei and their electrons is about:

$$V = 2 \times 10^{-44} \text{ m}^3.$$

So the empty space within each copper atom is the difference between 13×10^{-30} m^3 and 2×10^{-44} m^3. This give a ratio of the empty space to the volume of the nucleus and the electrons of 650 trillion to 1 and means that if the nucleus and the electrons were solid the percentage of empty space in a copper atom would be:

Percent Empty Space = 99.99999999999985 %

But according to the Ross Model the electron is not a solid; the naked electron is three circling point particles as shown in FIG. 5 and those point particles occupy no volume at all. The one conduction electron in the copper atom is an energetic electron which includes an entron but the entron is comprised of two extra point particles but they also occupy no volume. So each of the electrons in the copper atom is 100 percent empty space.

There are two stable isotopes of copper, Cu^{63} and Cu^{65}. The nucleus of the Cu^{63} isotope contains 63 protons and 34 electrons and a few entrons and the Cu^{65} isotope contains 65 protons and 36 electrons and a few entrons, according to the Ross Model. As explained in **Chapter VI** protons are comprised of electrons, positrons and entrons all of which are comprised of nothing but tronnies that occupy no volume. Therefore, the nucleus of the copper atom is 100 percent empty space. As a result we are left with the conclusion that instead of the copper atom being 99.99999999999985 percent empty space, it is 100 percent empty space! Since each atom in the 8 cubic centimeter cube of copper is 100 percent empty space, the 8 cubic centimeter cube of copper is 100 percent empty space. However, each tronnie in the 600 billion trillion atoms in the cube of copper is producing speed of light Coulomb force waves, so the empty space within the copper cube is completely filled with Coulomb force waves.

So our conclusion is, this 8 cubic centimetre copper cube is 100 percent empty space but that space is completely filled with speed of light Coulomb waves and believe it or not, at least according to the Ross Model, the same conclusion applies to the rest of our Universe, including you and me.

The Ross Model provides the first good description of charge. Charge is provided by tronnies and only tronnies. Tronnies get their charge from the Coulomb grid in which they are located. Much (maybe all) of the tronnie's charge comes from itself.

By traveling in a perfect circle at $(\pi/2)c$, the tronnie is always at the focus of a ring of charge produced by itself. The direction of the Coulomb force at the location of the tronnie changes by 360 degrees each cycle.

Our Point of View

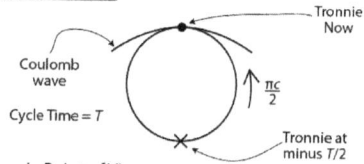

Coulomb wave

Cycle Time = T

$\frac{\pi c}{2}$

Tronnie Now

Tronnie at minus $T/2$

Tronnie Point of View

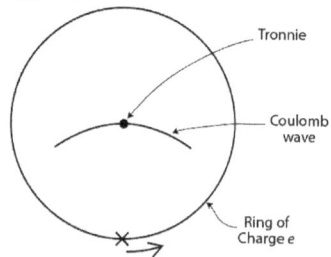

Tronnie

Coulomb wave

Ring of Charge e

CHAPTER XXIV

WHERE DO TRONNIES GET THEIR CHARGES?

What is Charge?

The Ross Model proposes that everything in our Universe is made from tronnies having no mass, no volume and a charge of plus or minus e. We all can image no mass and we all can image no volume. But what is "charge" anyway? And more specifically what is a charge of e. Scientists tell us that "charge" is a property of things such as electrons and protons that have a "charge". They know that the electron has a charge which they define as minus e (about minus 1.602×10^{-19} coulombs) and they know that the proton and the positron have charges of plus e. They also know that charges produce a force, the Coulomb force, either attractive or repulsive on things that also have charge. But scientists do not know what charge is or have a good explanation as to how the electrons, positrons and protons acquired their charges. If you don't believe me, GOOGLE "charge", "electric charge" or "electron charge".

The Tronnie's Charge

As you the reader are very well aware by now that the Ross Model proposes that all charges are carried by tronnies, in fact I think we could say that each tronnie **is** a charge since charge is the only property it possesses. It has no mass and no volume, only charge. Furthermore, that charge is exactly constant never diminishing, never increasing even though each tronnie is continually utilizing that charge to propel itself faster than the speed of light and in most cases (if not all cases) also to help propel other tronnies or other particles. So the question for this section is: "Where do tronnies get their charge, and how come it never decreases?"

The Ross Model proposes that each and every tronnie gets its charge from the Coulomb grid in which it is located or through which it is traveling. Remember every tronnie is always traveling at the speed of light or faster. (This current version of the Ross Model proposes that each tronnie travels in a circle at a speed of $\pi c/2$ with respect to at least one frame of reference.) It may be that each tronnie always travels in a perfect circle at the speed of $\pi c/2$. We learned in a previous chapter that our Universe is filled completely with Coulomb grids made up of Coulomb force waves all traveling at the speed of light. These speeds of light are all relative to some frame of reference. If you can imagine a 60 gram glass ball (silicon dioxide) the size of a baseball traveling at 100 miles per hour (about 44.7 m/s) relative to our Milky Way Galaxy in a direction away from our galaxy through interstellar space half way between our galaxy and our nearest galactic neighbour. That 60 gram glass ball contains about 6.022×10^{23} atoms of silicon and twice as many atoms of oxygen. Each atom of silicon contains 28 protons and 28 electrons (14 of which are in the nucleus) and each atom of oxygen contains 16 protons and 16 electrons (8 of which are in the nucleus). (In silicon dioxide each silicon atom transfers four of its electrons to the two oxygen atoms it is partnered with giving each oxygen atom two extra electrons.) Each electron contains at least 3 tronnies (2 minus tronnies and 1 plus tronnie) and each proton contains at least 11 tronnies, (6 plus tronnies and 5 minus tronnies). In addition the glass ball contains a bunch of entrons with two tronnies each, which we won't bother counting. So I think we have within this little glass ball more than 5×10^{26} charges each carrying a charge of plus or minus e and producing Coulomb force waves which are traveling through the glass ball and passing out from it through our Universe. According to my math 5×10^{26} is equal to 500 billion-trillion charges in this little glass ball. This is 500 billion-trillion tronnies each one repelling itself and all other like tronnies and attracting unlike tronnies. Furthermore, the combination of all of the Coulomb force waves (each traveling at the speed of light) from the tronnies (each moving faster than the speed of light) is creating a Coulomb grid within the ball and that grid is moving at 100 miles per hour through intergalactic space. We can think of a photon from our galaxy that may be approaching the ball at the speed of light relative to the universal Coulomb grid at that location in interstellar space in the same direction as the direction of travel of the ball. When the photon enters the ball it will adjust its speed so as to travel at the speed of light through the Coulomb grid within the glass ball. (My guess is that the Coulomb grid within the glass ball is dominated by the Coulomb waves produced by the charges within the glass, but there may be some significant contribution from the Coulomb grid at that location in the intergalactic space.) The photon's speed as it passes through the ball is determined by the index of refraction of the glass and the speed of the ball relative to interstellar space. The index of refraction of the glass relative to interstellar space will cause a decrease in the photon's speed but the ball's 100 mph speed in the direction of the photon will cause an increase in its speed. I have not figured out whether the photon speeds up or slows down as it passes through the ball. The point of this example is that the tronnies in the ball are creating a Coulomb grid that travels with the ball and controls or helps control the speed of light of any photons passing through the ball.

My other point that I am attempting to make here is that the Coulomb grid inside a little glass ball includes a tremendous number of Coulomb force waves all traveling at the speed of light. Now the questions is: Could all of these 500 billion-trillion charges packed into a tiny glass ball create a grid of forces moving at the speed of light that could focus themselves into 500 billion-trillion points (i.e. the tronnies in the glass ball) that are moving through our Universe at the speed of light and also circling at speeds faster than the speed of light. If so, we may have answered the question of where tronnies get their charge. The answer would be they get it from themselves and other tronnies.

In support of this proposition we may consider the entron in **FIG. 2B** in **Chapter IV**. In this thought experiment, let's think only in two dimensions as **FIG. 2B** portrays the entron. Let's assume the entron is a green light entron with a diameter of 3.77 X 10^{-10} m. The tronnies in a green light entron circle with a frequency of about 400 thousand trillion times per second. If you could imagine shrinking yourself down sufficiently so you could catch a ride on Tronnie P of the **FIG. 2B** entron at the time of the **FIG. 2B** snapshot you would be at location 200, and you would see Tronnie N at location 204 at the instant of the **FIG. 2B** snapshot, even though at that time Tronnie N is at location 202 on the opposite side of circle from Tronnie P. However, as you (riding on Tronnie P) circle along the circumference of the entron, you will perceive tronnie N making perfect circles around you at a distance of b = 0.59461d'. You would also see yourself and Tronnie P (that you are riding on) from your recent pass, circling you and Tronnie P at a distance of d'. (Remember if you are traveling in a circle at 1.57c, it is easy to see yourself from your past because the light from your past is traveling more slowly at a speed of c.) Since you and Tronnie P would be circling at 400 thousand trillion times per second; you and the tronnie you are riding on would appear like a ring with a radius equal to d' surrounding you and your tronnie. Tronnie N would also appear as a ring at a smaller radius equal to 0.59461d'.

The repulsive Coulomb forces from your tronnie would all be focused on your tronnie at all times sequentially at all points on the circular path of the tronnie and in all directions 400 thousand trillion times per second. As the focused integrated forces pass through the point that is your tronnie (tronnie P) they spread out in all directions (at least in the two-dimensional problem we are considering). These focused integrated forces spreading out in all directions from this single point (i.e. Tronnie P) moving at 1.57 c is at least a portion (maybe all) of the **"charge"** of Tronnie P.

Those of you who have studied physics may remember that the Coulomb force at the center of a static ring of charge is zero. However, the rings we are discussing are certainly not static rings of charge, each ring is the result of a single charge traveling faster than the speed of light.

A similar but much more complicated analysis can also be made with respect to the electron as shown in **FIG. 5** and the positron. As in the case of the entron, the tronnies of the electron are looping around each other in perfect circles faster than the speed of the Coulomb force waves so each one is continually receiving Coulomb force input from themselves and other tronnies. The charge of the electron is a **fact**. The electron must get its charge from somewhere. I say the electron gets its charge from the three tronnies that the electron is comprised of. I think there is a good probability that the tronnies of the electron (and the positron) receive their charges in a manner similar to the two tronnies in the entron. Keep in mind that the entron can have any size ranging from 0.934×10^{-18} m to a few centimeters. But the electron (and the positron) can exist in only a single size. It should be possible to create a computer model to analyse the Coulomb forces acting within the electron in order to determine if the tronnies of the electron can be provided with their charge from themselves only or from themselves and the other tronnies in the electron. (I hope to include a description of that computer model in the next version of this book. I am looking for a volunteer to help me with the model.) And it may be that the tronnies must receive some additional contribution from the Coulomb grid in which the electron exists.

Protons are merely combinations of electrons, positrons and entrons. It is therefore pretty clear to me that the proton gets its net charge from its two positrons and its single electron. Photons are made of entrons. Each photon and its entron have a zero net charge.

Since I cannot think of any other way that a tronnie could get its charge (a charge that never diminishes), at least for now, I propose that tronnies obtain their charge from themselves or from themselves and other tronnies and that each tronnie always moves along paths through our Universe such that it receives the precise amount of Coulomb forces from itself (and possibly other tronnies) to sustain its charge at a charge of plus or minus e.

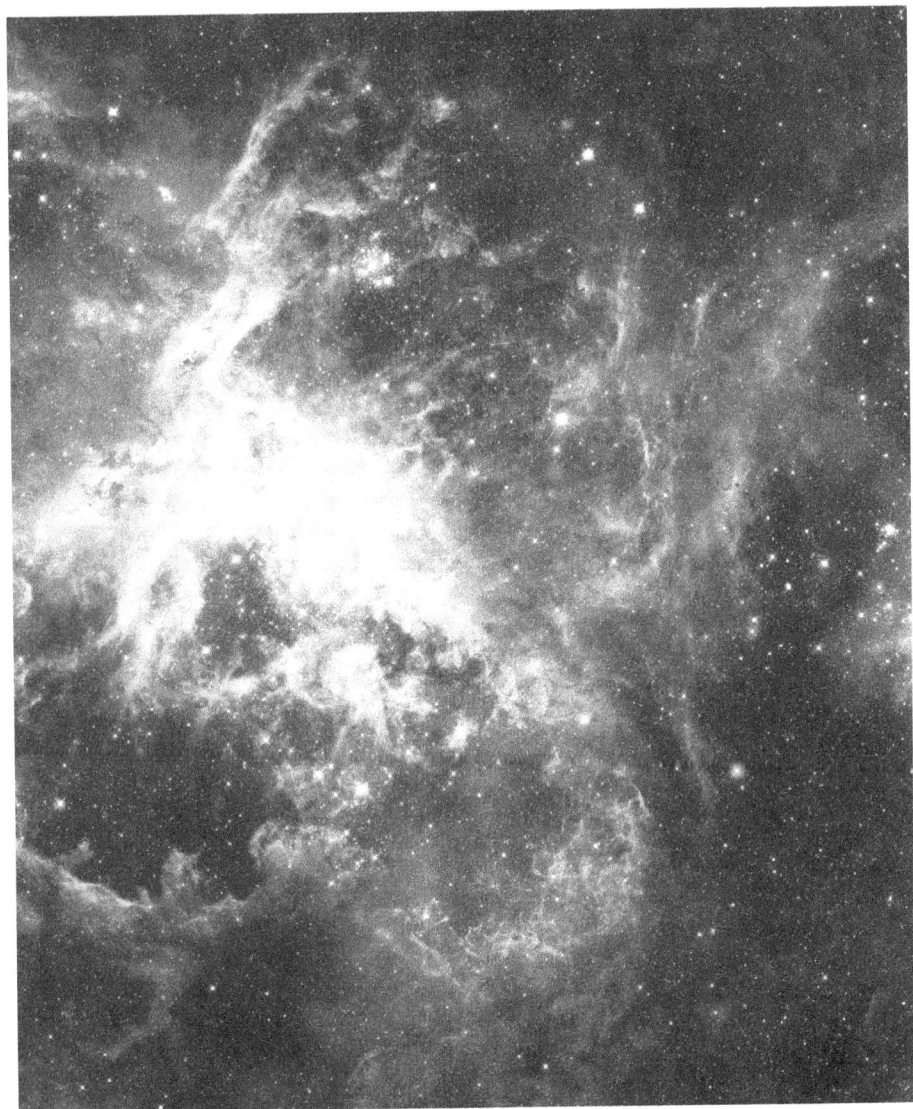

Tarantula Nebula

Image credit: X-ray: NASA/CXC/PSU/L.Townsley et al.; Optical: NASA/STScI; Infrared: NASA/JPL/PSU/L.Townsley et al.

The Big Bang that was the birth of our Universe was the death of our predecessor universe.

Our Universe has been expanding for about 13.7 billion years. The Ross Model speculates that after it has expanded for a total of about 50 billion years it will start to contract, and after about 50 billion years of contraction, almost all galaxies will be consumed in a single Monster Black Hole which will explode in a big bang explosion, marking the death of our Universe and the birth of our successor universe.

There is a substantial increase in the total mass-energy of each successive universe.

We can assume that our Universe may be the 47th universe and can extrapolate back to the very first tiny universe. And prior to the first tiny universe, there must have been a time period when and where the very first tronnies and/or the first Coulomb waves were created.

We do not know how the very first tronnies and/or the first Coulomb waves were created or who (if anyone) created them.

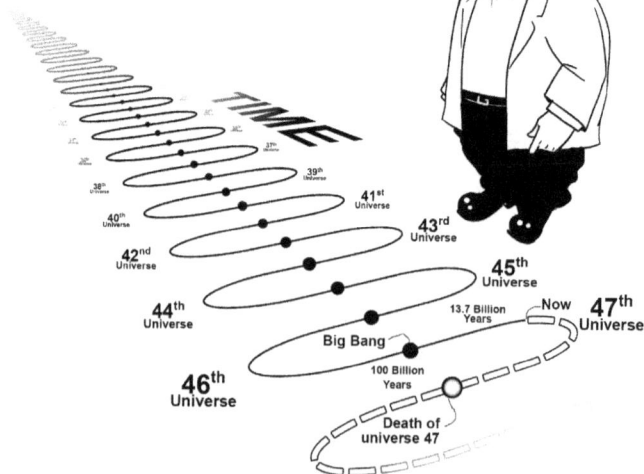

TIME

38th Universe
39th Universe
40th Universe
41st Universe
42nd Universe
43rd Universe
44th Universe
45th Universe
46th Universe
47th Universe

Now

13.7 Billion Years

Big Bang

100 Billion Years

Death of universe 47

CHAPTER XXV

LIFE AND DEATH OF UNIVERSES

Recycling on a Grand Scale

According to the Ross Model our Universe is a successor of a large series of universes. Each universe in the series is born in a Big Bang explosion that is the death of its predecessor universe. It develops a large number of galaxies, expands then contracts and ends its life in a Big Bang explosion which is the birth of its successor universe. This is recycling on a grand scale!

So, according to the Ross Model the universe we live in, our Universe, is a number in a series of universes. The current age of our Universe is about 13 to 15 billion years (I understand our best estimate is 13.7 billion years) and I estimate the lifetime of our Universe at about 100 billion years; so, according to my estimates, we have about 86 billion years to go. It is extremely unlikely that we will ever know what the number of our Universe really is. We could guess. For example, we might propose that our Universe is Universe Number 47, created at the demise of Universe Number 46 and when our Universe ends its life about 86.3 billion years from now, in the next Big Bang its recycled tronnies, along with perhaps some of the other debris of our Universe will create Universe Number 48.

The expansion and contraction of universes is caused by gravity and anti-gravity. Gravity and anti-gravity has been explained in **Chapters XXI and XXII**. The Ross Model attributes anti-gravity to the pressure produced by the impact of relatively low energy photon pressure exchanged between far away galaxies. Neutrino photons, the carriers of gravity within galaxies and among nearby galaxies, are absorbed and/or scattered by hydrogen atoms currently sparsely dispersed in inter-galactic space, so neutrino-photon-produced gravity is not at this time effective between very far-away galaxies. It is for this reason that our Universe is currently expanding.

The End of Our Universe

According to the Ross Model, when a sufficient portion of the free hydrogen in the space between galaxies has been pulled into galaxies and the stars of our Universe would have burned up most of the hydrogen in our Universe and would have become much dimmer than they now are, and the Black Holes would have become much larger and more concentrated. The neutrino photon flux from the Black Holes of the separate galaxies will then overcome the pressure of the remaining low-energy photons and begin to pull all galaxies together. In the course of this contraction, one gigantic Monster Black Hole near the center of our Universe will develop. This largest of all Black Holes will exist for a long time continuing to pull in surrounding galaxies including their Black Holes.

So I propose that our Universe will continue expanding for another 35 to 37 billion years when it will be about 50 billion years old. At that time it will begin to contract and the contraction will continue for another 50 billion years during which time all or almost all of our entire Universe (of about 100 to 400 billion galaxies), but excluding most of its shell, will be pulled into this Monster Black Hole. This means that on the average two to eight galaxies will be destroyed per earth-year. The rate of destruction will start slowly but the rate will increase until all or almost all galaxies in our Universe have been consumed. Since each galaxy contains on the average at least 100 billion star systems (including their associated planets and moons); at least about 270 million star systems will on the average be consumed per earth-day. That is equivalent to about 300 star systems per earth-second! Atoms will be broken down to protons, electrons and gamma ray entrons; then the protons will be destroyed to produce neutrino photons, electrons, positrons and gamma ray entrons as described in more detail in **Chapter XXI**. Nothing pulled into this Monster Black Hole escapes except neutrino photons. The resulting force of gravity produced by this flux of neutrino photons is sufficient to pull our entire Universe including the most distant galaxies toward the Monster Black Hole. Electrons and positrons will combine to produce gamma ray entrons. As our Universe contracts due to the increased production of the neutrino photons, the gravity between the galaxies and within galaxies will increase and the rate of contraction will increase.

When Worlds Collide at Very High Speeds

There is substantial support for the proposition that in a very short period, referred to as the inflation period, our Universe following the Big Bang expanded from a small volume to about 40 percent of its present size. If it did, the expanse must have been much faster than the speed of light. The Ross Model proposes that the Big Bang followed the collapse of our predecessor universe as has been explained above. Now I think I can also propose that the final collapse of our predecessor universe was also much faster than the speed of light, and that this much faster than the speed of light collapse may help explain the inflation period. Here are some simple calculations based on Isaac Newton's theory of universal gravitation that you readers should have learned about in your junior high school science classes.

His simple formula is:

$$F = GM_1M_2/r^2$$

where:

F = the attractive force of gravity between two objects, Object 1 and Object 2,
G is the gravitational constant (i.e. 6.67×10^{-11} m^3/kgs^2),
M_1 is the mass of Object 1 and M_2 is the mass of Object 2 and
r is the distance between the two objects.

Let M_1 be the equivalent mass of Object 1, the Monster Black Hole, and let Object 2 be a galaxy somewhere between about one light year (about 9.46 trillion km) from the Monster Black Hole and a location near the edge of our predecessor universe. Since force equals mass times acceleration, the gravitational acceleration acting on Object 2 is:

$$F = (M_2)a,$$

so the acceleration of M_2 is:

$$a = F/M_2 = GM_1M_2/M_2r^2$$

so:

$$a = GM_1/r^2$$

The acceleration of Object 2 does not depend on the mass of Object 2. (Remember, through empty space a cannon ball falls at the same rate as a feather. **So a galaxy should fall at the same rate as a cannon ball!**)

Our Universe has a mass of about 1.5×10^{53} kg. I have assumed that the Monster Black Hole during the last stages of the life of our predecessor universe had a mass equivalent to at least about 0.2 percent of the mass of our Universe or about 3×10^{50} kg. The distance r is the distance between the Monster Black Hole and Object 2 (i.e., anything falling into the monster Black Hole). The radius of universes can be as large as about 4.6×10^{26} m. So I am going to attempt to make a rough guess at the gravitation acceleration of galaxies located at various distances from the Monster Black Hole. I will assume that during the last half of the life of our predecessor universe, gravity trumps anti-gravity and Isaac Newton's formula applies throughout the universe. (We are in effect assuming that absorption of neutrino photons in the space between the Monster Black Hole and the galaxies can be neglected.) Since G is 6.67×10^{-11} m^3/kgs^2 and M_1 is 3×10^{50} kg, (the mass of Object 2 is irrelevant), the acceleration of Object 2 is:

$$a = (6.67 \times 10^{-11} m^3/kgs^2)(3 \times 10^{50} kg)/r^2$$

One light year is about 0.94×10^{17} meters (call it about 1×10^{17} meters). So let's estimate the acceleration in increments in the range of about 1×10^{15} m and about 1×10^{27} m.

After that, let us estimate the final velocities at 13 increments between these distances from the Monster Black Hole. Let's assume, to keep things simple, that the initial velocity at each increment is zero.

Final velocity v_f equals acceleration a (estimated with the above formula) times time t if the initial velocity is zero (i.e. $v_f = at$). Distance traveled equals average velocity divided by time traveled (i.e. $d = vt$ so $t = d/v$). If the initial velocity is zero, we can simplify things by assuming that the average velocity is the final velocity divided by 2, (i.e. $v_f/2$) so we replace t with d/v and replace v with $v_f/2$, so we can make a very rough estimate the final velocity using: $v_f = 2ad/v_f$, so $v_f^2 = 2ad$ or:

$$v_f = \sqrt{2ad}.$$

The result is presented in the **Table V** below:

Table V
Galactic Velocity Near the Death of Universe 46

Position	Distance D r	Delta ΔD $d_{P(N)} - d_{P(N-1)}$	r^2	Acceleration a	Final Velocity $v_f = \sqrt{2ad}$	$v_f = \sqrt{2ad}$
P	MBH to P_n (meters)	(meters)	(m^2)	$G(3 \times 10^{50}\text{kg})/r^2$ (m/s^2)	$v_f = \sqrt{2ad}$ (m/s)	$v_f = \sqrt{2ad}$ (v_f/c)
0	10^{14}					
1	10^{15}	0.9×10^{15}	10^{30}	2×10^{10}	6×10^{12}	2×10^4 c
2 (~1 ly)	10^{16}	0.9×10^{16}	10^{32}	2×10^{8}	1.9×10^{12}	6.3×10^3 c
3	10^{17}	0.9×10^{17}	10^{34}	2×10^{6}	6×10^{11}	2×10^3 c
4	10^{18}	0.9×10^{18}	10^{36}	2×10^{4}	1.9×10^{11}	6.3×10^2 c
5	10^{19}	0.9×10^{19}	10^{38}	2×10^{2}	6×10^{10}	2×10^2 c
6	10^{20}	0.9×10^{20}	10^{40}	2×10^{0}	1.9×10^{10}	6.3×10^1 c
7	10^{21}	0.9×10^{21}	10^{42}	2×10^{-2}	6×10^{9}	2×10^1 c
8	10^{22}	0.9×10^{22}	10^{44}	2×10^{-4}	1.9×10^{9}	6.3×10^0 c
9	10^{23}	0.9×10^{23}	10^{46}	2×10^{-6}	6×10^{8}	2×10^0 c
10	10^{24}	0.9×10^{24}	10^{48}	2×10^{-8}	1.9×10^{8}	6.3×10^{-1} c
11	10^{25}	0.9×10^{25}	10^{50}	2×10^{-10}	6×10^{7}	2×10^{-1} c
12 (~10^{10} ly)	10^{26}	0.9×10^{26}	10^{52}	2×10^{-12}	1.9×10^{7}	6.3×10^{-2} c
13	10^{27}	0.9×10^{27}	10^{54}	2×10^{-14}	6×10^{6}	2×10^{-2} c

So a body that is initially stationary, falling from 1×10^{16} m to 1×10^{15} m (which is a distance of roughly 9×10^{15} m, about nine thousand trillion meters) toward the Monster Black Hole would arrive at 1×10^{15} m (one thousand trillion meters from the Monster

Black Hole) with a velocity many thousand times the speed of light. I have not made a more precise estimate of the velocity. I will leave that job to people better at this type of math than me. (I wish Isacc Newton were still around.)

However we can think about a galaxy, initially at the edge of our predessor universe, and subjected to the contiuous gravitational acceleration from the Monster Black Hole, starting very slowly but increasing its speed each second for many billions of years. Let us also assume that the Big Bang occurred within a few earth hours or weeks prior to the arrival of our galaxy. My guess is that by the time the galaxy reached the vicinity of the Monster Black Hole its speed would have been somewhere between 5,000 and 100,000 times the speed of light! If we pick a number 50,000 times the speed of light (15 X 10^{12} m/s) and assume a distance of 15 trillion meters for the vicinity of the Monster Black Hole, the galaxy would pass through that distance in one second. After passing through the vicinity of the Monster Black Hole, this galaxies and all other similarily situated galaxies would have expanded out from the vicinity of the Monster Black Hole at speeds many thousand times the speed of light to produce the inflation period of our Universe.

(Many readers are going to be disturbed by my suggestions of galaxies falling toward the Monster Black Hole at speeds 50,000 times the speed of light, since they have been taught that nothing can go faster than the speed of light. All I can say to such objections is that these readers were also taught Isaac Newton's formulas for gravitational force and acceleration. Isaac Newton's formulas did not include a speed limit. Some may also claim that space could be warped near the Monster Black Hole. The Ross Model proposes that space is nothing and that space cannot be warped. One nice thing about the Ross Model is that it provides a logical explanation of gravity that does not require a concept of large masses warping space.

I will have more speculation about the contraction of our predecessor universe, the Big Bang and the inflation of our Universe in the following sections.

What Happened to All of That Kinetic Energy?

We saw in the previous section that much faster than the speed-of-light collapse of our predecessor universe might help explain the Big Bang and the inflation period of our Universe. But could it also help explain how our Universe could have gotten so big. Let us make an estimate of the energy-mass added to the Monster Black Hole based on an assumption that 200,000 galaxies of our predecessor universe crashed into the Monster Black Hole at a speed of 200 times the speed of light. Let us let the mass of a single typical galaxy be the mass of the Milky Way Galaxy = 2 X 10^{42} kg.

So the kinetic energy KE added to the Monster Black Hole from this single galaxy would be:

$$KE = (½)mv^2 = (½)m(200c)^2 = (1/2)(2 \times 10^{42} \text{ kg})(4 \times 10^4)(3 \times 10^8 \text{ m/s})^2$$
$$KE = 36 \times 10^{62} \text{ J}$$

Now we convert this energy into mass using $E = mc^2$, so the mass equivalent of the kinetic energy of a single typical galaxy would be about:

$$M = E/c^2 = 36 \times 10^{62}/9 \times 10^{16} = 4 \times 10^{46} \text{ kg.}$$

If we assume that our predecessor universe had within it 50 billion galaxys and was about one half as massive (7.3×10^{52} kg) as our Universe (1.46×10^{53} kg) and that 0.4 percent (2 million galaxies) crashed into the Monster Black Hole at speeds of 200 times the speed of light, the total mass M_t equivalent to their kinetic energy would be:

$$M_t = (2 \times 10^6) \times (4 \times 10^{46} \text{ kg} = 8 \times 10^{52} \text{ kg,}$$

which would be enough additional mass to slightly more than double the mass of our predecessor universe.

Can the Collapse of Universe 46 Explain Inflation of Universe 47?
A big question arises as a result of the above analysis; that is, can we explain the much faster than the speed of light inflation of our Universe by the much faster than the speed of light collapse of our predecessor universe?

If you believe that our Universe was born with a Big Bang event which was the death of our predecessor universe, and if we believe that the collapse of our predecessor universe was the result of gravity originating from a Monster Black Hole, then it should be easy for us to believe that just prior to the Big Bang millions or (maybe billions) of galaxies were diving toward the Monster Black Hole from all directions at speeds many thousand times the speed of light. During the late stages of our predecessor universe, the gravity of all galaxies would have increased by many orders of magnitude due to the greatly increased neutrino photon flux produced by the Monster Black Hole. In addition during this period there would have been a great consolidation of galaxies and all of the galaxies would have shrunk in size by orders of magnitude and their gravity would have increased by orders of magnitude.

The Ross Model proposes that during these last stages of our predecessor universe the Monster Black Hole and all of the galaxies accelerating toward it were enormously hot and the Black Holes at the center of each of the galaxies were all busy consuming their galaxies, destroying protons and releasing neutrino photons which were greatly increasing the gravity of our predecessor universe. These neutrino photons were directed in all directions ultimately reaching the mostly electron-positron shell of our predecessor universe. In a following section of this chapter, "Incubation vs. Inflation" explains how these neutrino photons combine with electrons and positrons in the shell of our predecessor universe to produce naked protons in a very large region of the shell which region could be as large as 40 percent of the present size of our Universe. But high energy entrons are then needed to convert the naked protons into high-energy, low-speed protons which constitute the nuclei of hydrogen atoms.

Dissecting the Big Bang

Two questions that arise from this analysis are: "When in this process did the Big Bang occur and how long did it take? Current explanations of the Big Bang generally propose that the Big Bang lasted only a very short period of time. Most of the stars, planets and moons of the galaxies being pulled into the Monster Black Hole near the end of life of our predecessor universe will have been consumed by the Black Holes at the center of each of the galaxies. These Black Holes will have been destroying protons and producing neutrino photons and providing the gravity for its galaxy. Millions or possibly billions of these sources of gravity are now descending, much faster than the speed of light, and piling up on the Monster Black Hole which has by this time shed much of its mass to create the gravity pulling in the incoming galaxies. The incoming galaxies are crashing into the Monster Black Hole at ever increasing speeds of many thousand times the speed of light, and they are all coming from different distances in the universe, so they are not all going to get there at the same time. It seems much more likely that at some point in the process, as the last of the galaxies are arriving, the combined masses of the incoming galaxies will dominate the Monster Black Hole, and the Monster Black Hole will lose its ability to contain its trapped energy. I propose that the Big Bang occurred before all of the galaxies of our predecessor universe have been consumed. Therefore, a small but significant portion of the galaxies of our predecessor universe might have been rapidly bearing down on the Monster Black Hole from all directions at speeds many thousand times the speed of light at the moment of the Big. Bang.

What Caused the Monster Black Hole to Explode?

A good question is what could cause the Monster Black Hole to suddenly explode in the Big Bang after operating as a black hole for almost 100 billion years without exploding? According to the Ross Model each Black Hole produces its gravity by destroying protons to release their neutrino entron. The process of destroying protons is normally very slow and complicated which could explain why Black Holes can be stable for billions of years as they consume portions of their own galaxy and in many cases surrounding galaxies. In order to destroy a proton, the Black Hole must first create an anti-proton which quickly combines with a proton and both are annihilated releasing two neutrino photons. To create an anti-proton the Black Hole needs two electrons, one positron and a neutrino entron. The Black Hole has plenty of neutrino entrons. Two electrons are no problem since each hydrogen atom consumed by the Black Hole has one of them associated with its proton. But free positrons are relatively rare in Black Holes. However as explained in **Chapter VI** the Black Hole can produce its own positions by the combination of three photons of proper energies. These photons are:

1) a 928 MeV neutrino entrom having a photon circle with a diameter of 0.85×10^{-15} m and an entron with a diameter of 0.934×10^{-18} m,

2) a 1.02 MeV gamma ray photon with a diameter of 7.7×10^{-13} m but its entron has a diameter of 0.85×190^{-15} m, and

3) a low-energy 1.12 KeV photon with its entron having a diameter of 7.7×10^{-13} m.

There are an enormous number of neutrino photons in all Black Holes which are produced in abundance with the destruction of protons. The 1.12 KeV entrons are also in abundance since this entron at a temperature of 2.16 million degrees Kelvin is in the middle of the temperature spectrum of most stars at about 2.61 million degrees Kelvin which are the food stuff of Black Holes. However the 1.02 MeV entrons are much rarer in Black Holes since these entrons correspond to a temperature about one thousand times higher at 2.38 billion degrees Kelvin. Thus, the relative rarity of these 1.02 MeV photons limit the electron-positron pair production in Black Holes and the corresponding rarity of positrons limits the production of anti-protons which imposes a limit of proton destruction in Black Holes and prevents runaway reactions and allows Black Holes to operate with stability for billions of years. However at temperatures approaching the range of 2.38×10^{9} K (as indicated in **Table V** in **Chapter V**) creation of 1.02 MeV entrons will begin to increase rapidly providing the missing ingredient for runaway production of anti-protons.

Thus, when the temperature of the Monster Black Hole reaches a temperature in the range of 2.38×10^{9} K and higher the pair production rate will begin to escalate rapidly, greatly increasing the rate of anti-proton production and proton destruction that will correspondingly increase the gravitational attraction of the Monster Black Hole further increasing the rate at which galaxies are crashing into the Monster Black Hole. At this point the velocity of the galaxies after continuous acceleration for many billions of years is now several thousand times the speed of light. Thus, within a few minutes or hours a significant portion of the entire universe is crashing into the Monster Black Hole, compressing it, increasing its temperature, and tremendously increasing its destruction of protons production and release of neutrino entrons which further increases its gravity.

With the destruction of most of its remaining protons the Monster Black Hole now is comprised of a relatively small quantity of protons and perhaps also a similar small quantity of atoms that have not been broken down to protons, but it is at this time containing a very large portion of the mass of its universe in the form of entrons divided among neutrino entrons and lower energy entrons mostly gamma ray entrons, x-ray entrons, ultraviolet entrons and visible light entrons. Almost all of these entrons have been bottled up for billions of years looking for a way to escape as photons. At some point the Monster Black Hole explodes in a Big Bang explosion releasing in the form of radiation what is left of our predecessor universe.

But now let us think about the incoming galaxies, which may be thousands of galaxies (many the size of the Milky Way Galaxy or larger) descending on the Monster Black Hole all within a time of a few minutes and each at thousands of times the speed of light. At these speeds, such as about 19×10^{12} m/s (even if the size of the exploding Monster Black Hole is a light year [about 10^{16} m] in diameter), the Galaxies with their tremendous inertia would pass through the Monster Black Hole in about 1,670 seconds (about one-half hour).

The Big Bang explosion of the Monster Black Hole would have released a gigantic quantity of energy mostly in the form of high-energy radiation. That radiation once released would spread out from the region of the Big Bang in all directions at the speed of light. However, the explosion would not have stopped the incoming galaxies. The inertia of these galaxies coming in at many times the speed of light would have carried them through the site of the explosion site with little reduction in momentum. So the remains of thousands or millions of incoming galaxies (including their black holes, stars, planets and moons) are now exiting the site of the Big Bang explosion at speeds of many thousand times the speed of light. The radiation released by the Big Bang explosion spreads out in all directions at the speed of light but the photons are traveling through galaxies that are spreading out from the region of the Big Bang explosion at speeds many times the speed of light so the radiation actual speed would be the sum of speed of the radiation plus the speed of the galaxies through which the radiation is moving. So, if a galaxy is exiting the site of the Big Bang at 20,000 c (20,000 times the speed of light), the Big Bang radiation passing through it would be exiting at a speed of 20,001 c. This, as far as I know is the first logical explanation of the inflation of our Universe that occurred at the birth of our Universe. At least one other possible explanation of inflation is incubation in the shell of our predecessor universe as described in the next section.

Incubation in the Shell

In addition to the idea laid out above, (i.e. that faster than the speed of light contraction could produce faster than the speed of light expansion); the Ross Model proposes that as our predecessor universe was being consumed; our Universe was being incubated in the shell of our predecessor universe. Most of these neutrino photons produced in the Black Holes would have ultimately ended up in the shell of our predecessor universe, many of which would have traveled distances of many light years through the inner regions of the shell. Remember from **Chapter VIII** protons are made from the combination of a neutrino photon one electron and two positrons. The shell is primarily comprised of electrons and positrons with a relatively few protons; therefore while our predecessor universe was being contracted and consumed in Black Holes including the Monster Black Hole, our Universe was being **incubated** in the shell of our predecessor universe with the creation of new protons. But the new protons are mostly naked protons and as explained in **Chapter VIII**, naked protons travel at 4.02×10^7 m/s and they are traveling too fast to form hydrogen atoms. They slow down by capturing gamma rays but until the Big Bang, there are relatively few gamma rays at distances deep in the shell.

The Big Bang would have released all of the mass-energy of our predecessor universe that was left in the Monster Black Hole at the time of the Big Bang event. Keep in mind, however, that the great majority of the mass-energy has at the time of the Big Bang already been released in the form of neutrino photons which have for about 50 billion been driving galactic matter into the Monster Black Hole and infiltrating the shell of our predecessor universe. All or almost all of these neutrino photons are distributed throughout a huge thickness of the remaining shell of our predecessor universe. I estimate this thickness to be about one half to two thirds the size of our existing Universe, not including its shell. (As explained in **Chapter XXII**, protons in the inner portion of our shell are currently capturing enough entrons to slow down sufficiently to allow them to capture electrons to become hydrogen atoms.) As explained above the Monster Black Hole at the time of the Big Bang consists primarily of entrons including high energy gamma ray entrons and neutrino entrons. These spread out through the inner portion of the shell, which is left over from our predecessor Universe, to provide the gamma rays needed to slow down the naked protons so they can form hydrogen atoms. As the naked protons slow down to speeds close to zero, each proton collects a naked electron create a neutral hydrogen atom. This changes the new universe from a universe of ions, mostly naked protons, naked positrons and naked electrons, and high-energy entrons to a universe of neutral hydrogen atoms and low-energy entrons and photons. Low-energy entrons travel easily as low-energy photons through neutral hydrogen but are absorbed and reflected by ions in the shell of universes. These reflected low-energy photons after billions of years create the cosmic background radiation we now measure sometime as static on our TV monitors.

The Expansion of Our Universe

Inflation

Following the Big Bang a large majority of what was the shell of our predecessor universe is not destroyed. The inner portion of the shell, many light years thick, contains a large number of neutrino photons and protons and a smaller number of gamma ray photons. These gamma ray photons from the Big Bang are absorbed by the naked protons which slows them down so they can capture an electron to become a neutral hydrogen atom. This permits the gamma ray photons to expand more quickly through the shell. The gamma rays photons are quickly absorbed by electrons, positrons and protons, many of the gamma ray photons will combine with a neutrino photon and a low-energy photon to produce additional electron-positron pairs as explained in **Chapter VI**. Some are absorbed by naked protons to produce neutral hydrogen atoms. And some eventually lose their high-energy through scattering events with electrons, positrons, protons and hydrogen atoms. As the gamma ray photons spread out through the shell and are absorbed or lose their energy the effect is a cooling and expansion of the new universe.

Production of Additional Electrons and Positrons

In the new universe there is a tremendous abundance of neutrino entrons, gamma ray entrons and lower energy entrons now populating our successor universe. These combine in pair production processes to produce additional electrons and positrons.

Production of Protons

The electrons can capture neutrino photons to become the very high-energy electrons needed to form protons. The high-energy electrons then capture two positrons to form naked protons. These naked protons capture entrons having combined energies of about 8.37 MeV to slow down to speeds close to zero. The protons then capture one electron each to form hydrogen atoms.

Formation of Stars

Lower energy photons provide an expansion force and the neutrino photons left over from the prior universe provide a gravitational force. The neutrino photon contraction force is greater than the expansion force in various regions of space so that matter in the new universe (nearly all of which is protons and hydrogen atoms) tends to congregate in those regions. At great congregations lower energy photons are absorbed and neutrino photons are absorbed and released randomly out of the congregation providing a type of gravity for the congregation producing further congregation of the congregations. These congregations become stars in which hydrogen is converted to helium releasing fusion energy in the form of gamma ray radiation and the hydrogen and helium are converted into larger atoms releasing additional radiation. Stars explode releasing atoms which form themselves into molecules and these molecules congregate into planets and moons.

171

Formation of Galaxies

Some stars become very large and collapse into Black Holes which begin destroying protons to produce neutrino photons and the resulting gravity. (In addition a few Black Holes from our predecessor universe may have survived the Big Bang explosion of the Monster Black Hole.) Nearby stars and their planets and the moons of the planets will be consumed by the Black Holes. Each proton destruction releases a neutrino photon to provide additional gravity effects. Fast moving close-by stars and far-away slower moving stars begin to circle the Black Hole. Stars generate larger atoms in fusion processes and heavier atoms are produced in explosions of stars. Planets are formed from the dust of these exploding stars. Dust from the exploding stars will include all of the 92 chemical elements present in our Universe including oxygen, hydrogen and carbon. So we expect that at least a few of these planets will orbit a star at a distance such that liquid water and carbon compounds will be present in abundance. Life will evolve with varying degrees of intelligence. A single form of life with the greatest intelligence will ultimately dominate the other forms of life. That life form will develop civilizations and several thousand years after civilizations develop some of these creatures will attempt to figure out how their universe was created. Religions develop from these early attempts which will provide some answers, most of which will be wrong. Ultimately one of these creatures will develop a **theory of everything** which will provide the correct explanation.

The Very First Universe

So if our Universe is Universe Number 47 created from the recycling of Universe Number 46 and Universe Number 46 resulted from Universe Number 45 and so forth back to the destruction of Universe Number 1, the obvious question is, "Where did Universe Number 1 come from?" My answer is: "I don't know."

However I believe that at some time in the extremely distant past, before there was a Universe Number 1, there were electrons and positrons, and before there were electrons and positrons, there were tronnies and Coulomb forces; but **sometime before there was anything, there was nothing, just empty space**. I also believe that tronnies are the focus of Coulomb force waves and that tronnies produce Coulomb force waves. But something must have started the process. I believe tronnies are created in pairs (one plus and one minus) and that a plus tronnie cannot be created without also creating a minus tronnie. My guess is that it is not possible to destroy a tronnie once it is created, and that at some time or times in the life of each universe (such as at the time of the Big Bangs or maybe in Black Holes or even in supernovas) a great many new tronnies are created so that successive Universes get larger and larger probably exponentially. Our Universe right now today contains 100 to 400 billion galaxies, each with a huge number of star systems.

To create the first Universe all we need are tronnies or positive and negative Coulomb waves traveling at a constant speed relative to some reference system that can be focused to positive and negative points. Once the positive and negative points are created they will naturally be attracted to each other and will tend to form entron circles. Interactions among the circles will create entrons of many sizes. My guess is that we will find that there will be a smallest possible circle with a diameter related to the speed of the Coulomb waves. I will also guess that if the wave speed is $c = 3 \times 10^8$ m/s the smallest possible circle is 0.934×10^{-18} m (i.e. the neutrino entron in our system).

Once we have entrons we have mass and once we have neutrino entrons, we have gravity to pull things together, and we have the building blocks to produce electrons and positrons and once we have electrons and positrons and neutrino entrons, we have the building blocks to produce protons. Once we have the protons, electrons and entrons we have the building blocks to make hydrogen atoms. And once we have hydrogen atoms and entrons we have the building blocks to make alpha particles and once we have alpha particles, protons, and electrons, we have the building blocks to produce all varieties of the 92 atoms from hydrogen to uranium. With atoms we can make molecules and with atoms and molecules we can make universes. So now the really big question is: **"Could all of this happen naturally or do we need an intelligent helping hand at least to get things started?"** I do not have the answer. I know what I want the answer to be. For if an intelligent helping hand is required then we need to ask: **"Where did the helping hand come from?**

Thoughts about Space and Time

Space

According to the Ross Model space is nothing and is infinite; our Universe is approximately spherical and consists of about 100 to 400 galaxies and is currently expanding. In our Universe including its shell empty space is completely filled with speed of light Coulomb waves forming a large number of Coulomb grids through which light travels. The shell of our Universe is comprised of a cold plasma comprised of mostly self-propelled electrons and positrons. Our Universe including its shell is located in an infinite space. There may be other universes with their shells also located in the same infinite space. Spatial measurements are used to quantify how far apart objects are.

Time

Time is independent of space and is absolute and infinite in both the past and the future. Measurements of time are used to quantitatively compare intervals between or duration of events. There is no such thing as "space-time". Time in our Universe can be measured from the Big Bang or it could be measured from any other event occurring before or after the Big Bang such as the creation of the first universe or the creation of the first set of tronnies or the creation of the first Coulomb wave. We may have some difficulty determining or reasonably estimating when events prior to our Big Bang really occurred. But we could take a guess.

Any Chance of Surviving the Next Big Bang?

The Ross Model proposes that our Universe will be destroyed by a Big Bang explosion in about 86.3 billion years which explosion will mark death of our Universe and the birth of our successor universe. I suppose we could conclude that we should not worry too much about this because everyone now living on earth will be long dead; plus, I am told that our earth will be destroyed by our sun in about 5 billion years. Nevertheless it does not hurt to think about how our descendants might survive the destruction of the earth in the near term and the destruction of our Universe in the long term. It should be easy to survive the destruction of the earth. We have about 5 billion years to build space ships that could carry some survivors of our civilization to another planet located in a Goldilocks zone of a nearby star (close enough to and far enough from a star that water on the planet is liquid).

Surviving the destruction of our Universe will probably be much more difficult, but we have about 86.3 billion years to figure out how to do that. I see two possibilities assuming the Ross Model is a reasonably accurate description of our Universe.

Our descendants may be able to sit out the Big Bang deep into the shell of our Universe. They would have to wait several billion years for our successor universe to cool off enough for some stars and planets to form then find a nice planet in a Goldilocks zone to relocate to. (Thinking about this possibility has made me realize that there could be some survivors of our predecessor universe that is traveling through our Galaxy right now looking for our earth.)

Another possibility for our descendants might be to move to a galaxy at the far edge of our Universe. There is a possibility, as indicated earlier in this chapter that the Big Bang that will destroy our Universe will occur before all of the galaxies of our Universe have been consumed by the Monster Black Hole. In that case the remaining galaxies which at that time will be hurtling toward the exploding Monster Black Hole at speeds up to many thousand times the speed of light and that the inertia of the galaxies will carry them through the location of the explosion into the new universe. The neutrino flux would be enormous but life forms living in these galaxies may be able to survive the neutrino photon radiation as we now survive the much lower fluxes. The life forms may be totally unaware that they are traveling faster than the speed of light. And if at the time of the Big Bang, they are a few light years away from the site of the Big Bang they may not even know it had occurred when they pass trough the site.

174

Another possibility is to design a very fast space ship. Then our descendants would aim the spaceship toward the Monster Black Hole so that the gravity of the Monster Black Hole accelerates the space ship to many thousand times the speed of light but they would need to control the direction of the ship so that it makes a near miss of the Monster Black Hole a few days or weeks before the Big Bang. Inertia plus the ship's own power could then carry our descendants into the next universe. They would also have to guide the ship carefully to avoid the stars, planets and moons of the incoming galaxies on the opposite of the Monster Black Hole.

You the reader may have some additional thoughts of plots for science fiction books or movies where our decendents are in the process of trying to survive the destruction of our Universe.

This chapter provides a detailed description of the most important atom in our Universe, the hydrogen-1 atom. You have already learned in Chapter III how naked electrons combined with a single neutrino entron and two positrons to become high-speed protons. You also learned in Chapter III how these high-speed protons then captured gamma ray entrons during the early phase of our Universe to slow down enough to capture an electron to become hydrogen-1 atoms, and in the process helping to cool down our early Universe. You learned in Chapter XIII how stars make helium and all other atoms from electrons, positrons, protons and entrons while releasing energy carrying photons as a byproduct to provide heat and light to their surrounding planets.

You may know that at normal conditions hydrogen-1 atoms do not like to be alone and that they readily combine with other hydrogen-1 atoms to become hydrogen molecules. Lucky for us, two hydrogen atoms combine with an oxygen atom to become water, H_2O. Hydrogen atoms also combine with a great variety of other atoms and these combinations are essential in the chemical processes of life itself. In this Chapter XXVI we make a more detailed investigation of the hydrogen-1 atom, starting where Niels Bohr left off about 100 years ago.

CHAPTER XXVI

THE ROSS ATOM

The Bohr Atom

Almost exactly 100 years prior to my writing of this chapter Niels Bohr provided a description of atomic structures which was a great improvement over then existing descriptions of atoms and their structures. His atoms each had a massive positively charged nucleus surrounded by one or more orbiting electrons. He provided a detailed description of the simplest of all atoms, the hydrogen-1 atom which he described as a proton surrounded by a single electron. His model of the hydrogen-1 atom required the electron to be confined to certain specific orbits. His model required the electron in each of these specific orbits to have an angular momentum equal to an integer multiple of the lowest orbit. The electron could jump from orbit to orbit only by absorbing or emitting a specific quantum of energy. His model provided results that were in very good agreement with spectral data relating to ultra-violet, visible and infrared radiation absorbed and emitted from hydrogen gas, including hydrogen gas in our sun and the stars.

In the years since Professor Bohr introduced his model complicated mathematical descriptions of atoms including the hydrogen-1 atom and its electron have replaced the Bohr atom in scientific circles and the Bohr atom is considered a "relatively primitive model". The Ross Atom is a take-off from the Bohr Atom based on the Ross Model of our Universe and provides some additional support for the basic structure of the atom, its nucleus and its election as originally described by Niels Bohr. Professor Bohr was not aware of tronnies and entrons and the internal structure of electrons, but neither are the scientists that criticize his model. The Ross Atom is inconsistent with modern theories that describe the proton as being made of quarks and describe the electron as being a wave function. Under the Ross Model the nucleus of the hydrogen-1 atom is comprised of a proton that is in turn comprised of a massive high-energy electron and two positrons plus in most cases a variety of gamma ray entrons that are needed to slow down the proton to a speed close to zero. Under certain conditions the nucleus is orbited by or associated with a single electron which may be a very high-speed electron in a ground state or an excited state having captured one or more entrons to slow it down to certain specific velocities.

The Hydrogen Molecule

Most people when learning about atoms first consider the "Bohr Atom" which they learn is merely a proton associated with an electron. And they learn that the atom can have a variety of energy states defined by hydrogen spectral lines which were known at the time

Niels Bohr first described his atom. Many people would be surprised to learn that the Bohr atom is extremely rare or non-existent at normal temperatures and pressures. This is because at normal temperatures and pressures atomic hydrogen is very unstable. Its electron, as described by Professor Bohr is traveling alone in its circle around the proton (with a charge of +1) at speeds of about 2.19 million meters per second. The electron would much prefer to be circling an oxygen nucleus along with the electron of another hydrogen atom and oxygen's 8 regular electrons. So each of two hydrogen atoms gives their electron to the oxygen atom. This gives oxygen 10 electrons (2 in a first orbit and 8 in a second orbit) and a charge of minus 2. The two protons that have been abandoned by their electrons and are now positive ions. They are attracted to and attach themselves to the negatively charged oxygen atom to make charge neutral water, H_2O. That is why we have so much water here on earth. Hydrogen similarly forms bonds called covalent bonds with most other chemical elements. And it also forms covalent bonds with itself forming the H_2 molecule which exists as a solid at temperatures below 14.1 K, as a liquid above 14.1K and below 20.28 K and as a gas above 20.28 K. At temperatures a few 1,000 K the hydrogen atom exists as atomic hydrogen, and above about 31,700 K hydrogen is almost entirely in a plasma state, with the proton nucleus separated from its electron and with both being driven with energies in excess of 13.6 eV.

The Hydrogen-1 Atom
The Most Important Atom in Our Universe
The simple hydrogen-1 atom is the most important atom in our Universe. The nuclei of these tiny atoms carries within themselves all of the energy needed to fuel all of the stars (including our sun) in our Universe. And they carry within themselves all of the neutrino entrons needed to provide all of the gravity of every galaxy in our Universe.

The Hydrogen-1 Nucleus
Protons are described in detail in **Chapter VIII**. The nucleus of the hydrogen-1 atom is a very energetic proton, which is a proton that has been slowed down from its natural speed of 4.02×10^7 m/s (40.2 million meters per second) to a speed much closer to zero by the capture of about 8.37 MeV of gamma ray entrons. As explained in **Chapter VIII**, I am assuming that it takes about 15 gamma ray entrons each with energy of about 0.558 MeV to provide the total 8.37 MeV. Each of these about fifteen 0.558 MeV gamma ray entrons would have a diameter of about 1.555×10^{-15} m which is somewhat larger than the diameters of the circular paths of the electron and the two positrons in the proton. As you might suspect, I am merely guessing at the number, energy and size of the gamma ray entrons. I am somewhat more confident about the total energy of all of the entrons, and what I mean by "much closer to zero speed" is relative to 4.02×10^7 m/s, so in this context "close to zero" could still be pretty fast.

Once protons have been slowed down to "much closer to zero" speed their speed can be changed by the capture of additional entrons. These additional entrons can be provided by an electric field or heat energy. The temperature of an environment is a measure of the energy (or wavelengths) of the photons being absorbed and emitted by particles within the environment including the protons in the environment. When protons absorb a photon as explained in **Chapter VIII** the entron in the photon becomes an integral part of the proton and increases the mass and kinetic energy of the proton in accordance with m $=E/c^2$ and $E = (\frac{1}{2})mv^2$. In most cases the mass of the entron captured by protons is negligible compared to the proton mass except in the case of very high energy gamma ray entrons. (See **FIG. 8** in **Chapter IX**.) Photons with energies in the range of 10.2043 eV and 13.6057 eV can be captured by ground state electrons as described below without the electron being ejected from the hydrogen atom as explained in the next section.

The Orbiting Electron

In atomic hydrogen-1 the energetic, close to zero-speed proton may be orbited by a single electron. The electron may be in its ground (zero electrical energy) state in which case its velocity is about 2.1877×10^6 m/s (about 2.1877 million meters per second) and its stable orbit (as calculated by Niels Bohr) is one with a radius of about 5.29×10^{-11} m. This velocity of about 2.1877×10^6 m/s gives the ground state electron a kinetic energy E of about 13.6057 eV determined by $E = \frac{1}{2} mv^2$. At this speed and radius, the electron orbits the proton with a frequency of about 6.588×10^{15} cycles per second (about 6 ½ thousand-trillion cycles per second). Its angular momentum, mvr is about 10.545×10^{-35} kgm^2/s and the product of its velocity and radius is about 1.157×10^{-4} m^2/s.

Readers should keep in mind that in this section we are talking about the proton in atomic hydrogen. At normal temperatures, as explained above, atoms of hydrogen very quickly pair up to form molecular hydrogen in which case (according to the Ross Model) two atoms of hydrogen combine with the two protons circling a common center and the two electrons are circling through the circular path of the two protons to form the H$_2$ molecule. At very high temperatures such as at the surface of our sun where the temperature averages about 5,800 K, the environment is filled with photons with energies averaging about 2.48 eV which is in the green light portion of the visible spectrum, but the environment at the surface of our sun also includes photons with higher and much lower energies. The diameter of a 2.48 eV entron is about 3.5×10^{-10} m, larger but not much larger than the hydrogen-1 atoms. At these temperatures the protons of the hydrogen atoms capture entrons and acquire the thermal energy of the entrons (averaging about 2.48 eV, equivalent to about 3.97×10^{-19} J). Since $E = (\frac{1}{2}) mv^2$, this gives the protons a velocity averaging about 2.18×10^4 m/s (about 21,800 m/s). At these velocities the hydrogen molecule is not stable and splits into two protons and two electrons with the protons traveling at velocities of 2.18×10^4 m/s and the electrons traveling at velocities about 100 times faster. So hydrogen **molecules** do not exist and the hydrogen-1 nucleus may exist as an ion (separated from its electron) or it may exist as atomic hydrogen as described by Niels Bohr.

At the surface of our sun most of the electrons in most of the atoms of hydrogen will be in an excited state due to the capture of one or more entrons providing a total energy for each atom in the range of about 10.2043 eV to 13.6 eV. In the event the total energy of the entrons captured by the electron of a single atom exceeds 13.6 eV the electron will be left with a velocity that is too slow or too fast to support a stable orbit around the proton and the electron will become detached from its proton. A ground state electron of a hydrogen-1 atom cannot capture an entron with energy less than 10.2043 eV. This entron represents the lowest energy of the Lyman series and has a diameter according to the Ross Model of 0.849×10^{-10} m. A ground state electron in a hydrogen-1 atom orbits the proton at a radius of about 0.529×10^{-10} m as described by Niels Bohr. With the capture of the 10.2043 eV entron the electron is slowed down from 2.1877×10^6 m/s to 1.0938×10^6 m/s and its orbit is increased from 1.0548×10^{-10} to 4.2334×10^{-10} m. In addition the angular momentum of the electron is exactly doubled and the hydrogen atom is in a stable configuration as least temporally.

Hydrogen in Our Sun

Our sun is comprised of about 75 percent hydrogen (most of the rest being helium). The temperature of our sun at its core is about 15,700,000 K (about 15.7 million degrees Kelvin) and the temperature at its surface is much cooler, about 5,800 K. The absorption of energy of 1312 kJ/mole (equivalent to about 13.6 eV per atom and a temperature of about 31,200 K) will separate the hydrogen into an energetic proton and an energetic electron. So a very great majority of the hydrogen in our sun is in a plasma state. But at the surface most of the hydrogen is at a temperature in the range of about 5,800 K and most of the hydrogen is in its atomic form. Our sun is basically a black body radiator based primarily on its surface temperature of about 5,800 K. This produces a radiant energy distribution according to Wien's Law that peaks at about 2.48 eV (corresponding to a wavelength of about 5×10^{-7} m which is blue-green visible light, see **Table V** in **Chapter V**).

Atomic Hydrogen

As Niels Bohr explained about 100 years ago, the electron in the hydrogen-1 atom is stable in its ground state orbiting the proton at a velocity of about 2.1877×10^6 m/s at a radius of about 0.529×10^{-10} m. This velocity of about 2.19×10^6 m/s gives the electron a kinetic energy of about 13.6057 eV. Niels Bohr also explained that the hydrogen-1 atom can also be stable (at least temporally) with its electron in one of many specific excited states and energy is released when the electron in an excited state returns to a lower excited state or to the ground state. The stable states are those for which the angular momentum of the electron is an integral multiple of the ground state of the electron.

According to the Ross Model, these excited states result from the capture by the orbiting electron of one or more entrons having energies in the range of 0 eV to 13.6057 eV. At the surface of the sun there exists a tremendous flux of photons peaking at energies of about 2.48 eV with huge numbers of photons in the range from close to 0 eV to 13.6057 eV. Each photon is an entron traveling in a circle at a speed of 2c (twice the speed of light) and forward at a speed of c as explained in **Chapter V** (see FIG. 4). When the electron of the hydrogen-1 atom captures an entron, the two tronnies of the entron circle through the center of the electron, once each cycle of the entron. The two tronnies of the entron each carry a charge of e (about plus or minus 1.602×10^{-19} coulomb) and each of them apply a force on the electron on each pass through the electron. According to the Ross Model, in order for an atomic hydrogen atom to be stable the total energy of all of its captured entrons must be in the range of 10.2043 eV to 13.6057 eV. If the total entron energy in greater than 13.6057, the orbiting electron will undergo a change in direction, and the electron will be separated from the nucleus of the atom. Entrons with energies less than 10.2043 eV will have a diameter greater than 0.8491×10^{-10} m and cannot be captured by the ground state electron which is orbiting the proton at a radius smaller than the diameter of the entron. The total energy of all of the captured entrons range in energy at any particular time between about 10.2043 eV and 13.6057 eV which tends to slow down the electron from its natural velocity of 2.1877×10^6 m/s to certain specific velocities between 1.09×10^6 m/s and zero speed. The frequencies of the individual entrons are proportional to the energy of the entrons and range from close to zero to about 3.29×10^{15} cy/s. A single entron with a frequency of 3.29×10^{15} cy/s has energy of 13.6057 eV which is the energy needed to slow the electron down to a speed approximately equal to zero.

Electrons can capture more than one entron in which case their energies are additive. For example, if an electron captured a 10.2043 eV entron (with a frequency of 2.467×10^{15} cy/s) and also captured a 1.8896 eV entron (with a frequency of 0.8824×10^{15} cy/s), the number of tronnies passing through the center of the electron per second would be:

$$2 \times (2.467 \times 10^{15}/s + 0.4569 \times 10^{15}/s) = 5.848 \times 10^{15}/s.$$

The result is the same as if the electron had captures a single entron with energy of 12.0939 eV with an frequency of 2.9239×10^{15} cy/s. In either case the number of tronnies passing through the electron each second is the same (i.e. $5.848 \times 10^{15}/s$); the energy of the electron would be reduced from 13.6057 eV to 1.5118 eV; the electron

181

would be slowed down from 2.1877 X 10^6 m/s to 0.7292 X 10^6 m/s; and the orbit diameter of the electron would be increased from 1.0584 X 10^{-10} m to 9.528 X 10^{-10} m. Other combinations of entrons which are responsible for the Lyman series of spectral lines are shown in **Table XI-A**. Therefore, within the total energy range of about 10.02 eV to 13.6057 eV each entron captured slows down the electron by an amount determined by the energy of the entron. The resulting slower velocity will cause an increase in the orbit radius r_{or} of the electron in accordance with the following formula developed by Niels Bohr more than 100 years ago:

$$r_{or} = ke^2/mv^2 \qquad (6)$$

where r_{oe} is the orbit radius, k is Coulomb's constant (about 8.99 X 10^9 Nm^2/C^2), e is the electron charge (about 1.602 X 10^{-19} C), m is the mass of the electron (which is about 9.109 X 10^{-31} kg) and v is the velocity of the electron. (As an aside, the reader should note that the entron has mass and entrons captured by an electron adds mass to the electron in proportion to the energy of the entron in accordance with Professor Einstein's famous formula: $E = mc^2$. However with entron energies in the range of 0 eV to 13.6 eV (see **Table V**) the entron mass is so small compared to the natural mass of the electron (equivalent to about 0.51 million eV) that the entron mass can be ignored.)

For example, the capture by an electron in its ground state of a 13.6057 eV entron will reduce the kinetic energy of the electron to approximately zero and its corresponding orbit radius will be very large (effectively infinite) compared to its ground state radius of 0.529 X 10^{-10} m and the electron becomes dissociated from the proton. But the electron is now an excited electron with an electrical energy of about 13.6057 eV as the result of its captured 13.6057 eV entron. The electron could release its 13.6057 eV entron as a 13.6057 eV photon, regain its velocity of 2.19 X 10^6 m/s and its kinetic energy of about 13.6057 eV and return to its stable orbit around its old proton or any new proton not already having an orbiting electron.

Since the electron kinetic energy $E_k = (1/2)mv^2$, and $v^2 = 2E_k/m$:

$$r_{oe} = ke^2/2E_k$$

where E_k is the kinetic energy.

If we substitute values for Coulomb's constant and the electron charge in Equation (6) we can solve for r_{oe} in terms of the electron velocity and photon energy, and the orbit radius becomes:

$$r_{oe} = (2.532638003 \text{ X } 10^2 \text{ m}^3/\text{s}^2)/\text{v}^2$$

and

$$r_{oe} = 11.53539778 \text{ X } 10^{-29} \text{ Nm}^2/E_k$$

where E represents the kinetic energy of the electron and is in joules, r_{oe} is the electron orbit radius in meters and v is the electron velocity in meters per second, and

$$r_{oe} = 7.199825863 \text{ X } 10^{-10} \text{ eVm}/E_k$$

where E_k is in eV.

So if the electron is in its ground state with an energy of 13.6057 eV and a velocity of about 2.1877 X 10^6 m/s, its orbit radius will be about 0.5292 X 10^{-10} m and its orbit diameter will be about 1.0584 X 10^{-10} m. Since orbit frequency is f = v/πd, the electron will orbit the proton with a frequency of about 6.6540 X 10^{15} cycles per second. That is a little more than about 6.6 thousand-trillion cycles per second.

We see from the above that the radius r_{oe} of the orbiting electron is inversely proportional to the kinetic energy of the electron and specifically:

$$r_{oe} = 7.199825863 \text{ X } 10^{-10} \text{ eVm}/E_k$$

and the orbit diameter d_{oe} of the electron is twice as large at:

$$d_{oe} = 14.399651173 \text{ X } 10^{-10} \text{ eVm}/E = \text{about } 14.400 \text{ eVm}/E_k$$

where E_k is the kinetic energy in eV of the electron.

Photon Energy and Spectral Lines
We know that the wavelength λ of a photon is inversely proportional to the energy E_p of the photon and the specific relationship is:

$$E_p = hc/\lambda$$

where h is Plank's constant and c is the vacuum speed of light.

As explained in **Chapter V** the photon diameter d_p is related to the photon wavelength by:

$$\lambda = 2d_p/\pi = d_p/0.6366,$$

since (as explained in **Chapter V**) within the photon the entron is traveling in a circle at a speed of 2c while the circle is moving forward at a speed of c. The wavelength λ is the distance traveled forward by the entron during each cycle. If we replace λ with $d_p/0.6366$, we see that:

$$E_p = 0.6366 \ hc/d_p$$

and:

$$d_p = 0.6366 \ hc/E_p.$$

We get hc from **Table V** (i.e. hc = 12.39842435 X 10^{-7} eVm) so:

$$d_p = 7.892836941 \ X \ 10^{-7} \ eVm/E_p.$$

As explained in **Chapter V**, I have estimated that the ratio of the photon diameter d to its entron diameter d' is:

$$d_p/d' = \text{approximately } 911.$$

If this is correct:

$$d' = 8.66392639 \ X \ 10^{-10} \ eVm/E_p.$$

For example, the photon energy corresponding to the first spectral line of the Lyman Series is approximately 10.2043 eV (with λ = about 1.216 X 10^{-7} m), so the diameter of this photon would be 0.773481467 X 10^{-7} m and its entron diameter would be 0.849046616 X 10^{-10} m.

Examining Starlight

As explained above the temperature at the core of our sun is estimated to be about 15.7 million degrees Kelvin. At temperatures higher than 31,700 K the electrons are striped from the hydrogen-1 atom and the protons and electrons exist separate from each other in very hot plasma. The surface temperature our sun and most stars is much less than 31,700 K. Stars including our sun radiate light in the form of photons. The energy of the photons depends in large part on the temperature of the surface of the radiating star. The photon energy is distributed over a large range of energies, but if photon intensity is plotted against photon energy the peak intensity is relatively well consistent with Wien's Law referred to in **Chapter I**. The photon intensity at specific energy ranges will be lower at both higher and lower photon energies. According to Wien's Law:

$$E_p = 4.28 \times 10^{-4} \text{ eV-T/K}$$

Where E_p is the peak photon energy in a plot of photon energy vs intensity of the radiated photons and T is the star surface temperature in degrees Kelvin. For example, the surface temperature of our sun is only about 5,800 K and the light that we measure from the sun is determined by the surface temperature. The peak intensity of the solar spectrum is about 2.48 eV which is in the visible portion of the solar spectrum and is blue-green light. We use this value to determine that the surface temperature of our sun is about 5,800 K. The solar spectrum also includes the rest of the visible spectrum and the ultraviolet and infrared portions of the full electromagnetic spectrum. Other stars radiate light with a variety of photon energies depending primarily on their surface temperatures as explained above. Stars (including our sun) are surrounded by atmospheres and for most stars the atmospheres include hydrogen gas at a relative high temperature so that the hydrogen is in its atomic form (as opposed to earth's atmosphere where the hydrogen exists in its molecular form (H_2). As star radiation passes through star atmospheres, atoms in the atmospheres absorb specific frequencies of the light creating dark lines in the spectra of the light measured by astronomers here on earth. These spectral lines tell these astronomers most of what we know about the stars. These spectral lines have been known for more than 100 years. A very interesting thing about these spectral lines that our scientists see in starlight is that the opposite of the dark lines can be created here on earth by heating chemical elements to a temperature high enough to cause them to radiate. They radiate at energies that are the same as the energies they absorb in the atmospheres of stars. The spectra of the hydrogen-1 atom has been extremely well studied and is divided into a number of spectral series named after the discoverers of the various spectral series. These include the Lyman series, the Balmer series, the Paschen series, the Brackett series and the Pfund series and the Humphreys series. Other series are known but are unnamed.

All of the values of photon energies in the above series can be measured to a high level of accuracy and as explained above they have been known for about 100 years. A tremendous amount of scientific thinking has gone into attempting to explain why the hydrogen atom absorbs and radiates photons with precisely these energy values. To the best of my knowledge existing scientific theories do not provide a good explanation for why the lines are where they are. The Ross Model attempts to do that. You might ask, "Why do I think the Ross Model could provide this explanation when the smartest scientist in the world after more than 100 years of trying could not do it?" My answer is that those scientists have not known that orbiting electrons are self-propelled at speeds of about 2.1877 million meters per second and that they are slowed down by the capture of entrons having energies totaling between about 10.2043 eV and 13.6052 eV. In fact they have not even been aware of the existence of entrons. They knew that the lowest energy of the photon in the Lyman series had an energy of about 10.2043 eV and they were aware that the frequency of this photon (2.467×10^{15}/s) was in the same range as the orbit frequency (6.659×10^{15}/s) of the electron in the ground state hydrogen 1 atom, but they were not aware that its entron had a diameter (0.8491×10^{-10} m) very close to the orbit diameter of the electron in the ground state hydrogen atom (1.0584×10^{-10} m).

The Lyman Series

Theodore Lyman discovered the first line of the Lyman series in 1906 and additional lines of the series in the period 1906-1914. The spectral lines of the Lyman Series are described by the following formula:

$$E_p = (13.6056925 \text{ eV})(1 - 1/n^2) \quad (8)$$

where n is 2, 3, 4, 5, etc. and Ep is the photon or entron energy in electron-volts. These spectral lines are lines produced by spectrometers which are the instruments used to measure the spectral energy of photons. Measurements of spectral lines can be expressed in terms of electron-volts, joules, frequency and/or wavelengths. You can convert from one to the other using:

$$E_p = hc/\lambda = hf$$

where E_p is the photon energy in energy units, h is Plank's constant, λ is the photon wavelength and f is the photon frequency. Make sure you keep your units organized. Since $h = 4.1356692 \times 10^{-15}$ eVs,

$$f = E_p/4.1356692 \times 10^{-15} \text{eVs}$$

so for a 13.6057 eV photon, its frequency is 3.2898×10^{15} cy/s.

And since the entron diameter is:

$$d' = 8.66392639 \times 10^{-10} \text{ eVm}/E_p,$$

the diameter of the 13.6057 eV entron is about 0.6368×10^{-10} m (a little larger than one-half the orbit diameter of the hydrogen-1 electron in its ground state).

If you plug in the numbers for n into **Equation (8)**, you get the energy of the photons in the Lyman series. These energy values are listed in the second column of **Table XI-A**. If you subtract the energy corresponding to n from the energy corresponding to n+1 you get the energy difference in eV in column 3. This is the energy of the captured entron which will take the electron from state n to state n+1. I call this the "Delta Entron Energy Difference". We assume this is the energy of an entron captured by the electron in state n-1 to produce state n. So state 2 is created by the capture of a 10.2043 eV entron. This reduces the electron's velocity from 2.1877×10^6 m/s to 1.0938×10^6 m/s and reduces its kinetic energy to 3.4014 eV. This results in an increase in the electron orbit diameter to 4.2334×10^{-10} m. This decrease in the electron's velocity and corresponding increase in its orbit radius doubles the electrons angular momentum. This is a temporally stable orbit for the electron in the hydrogen-1 atom. This temporally stable orbit is the basis for the Balmer series of spectral lines. Two things can happen. The electron can lose its captured entron as a photon and return to its initial velocity and its ground state lower orbit. Or this 3.4014 eV electron can capture an additional entron. We will look at the Balmer series below, but for now let us continue with the Lyman series.

In **Table XI-A** and **Table XI -B** I have details regarding each of the Lyman lines from a state n = 1 to a state n = 10 and for a single very large n. The following describes the information provided in **Table XI-A**:

*Column 1 identifies Lyman lines 1 through 10 and lines 166 and 167.

*Column 2 is the total entron energy captured by the hydrogen 1 electron. (This could be a single entron or multiple entrons.)

*Column 3 is the energy a single entron to increase the photon energy from that of state n to that of state n+1.

*Column 4 is the electron kinetic energy for state n.

*Column 5 is the electron orbit diameter for state n.

*Column 6 is the electron velocity for state n.

*Column 7 is the entron diameter of an entron needed to move the electron from state n-1 to state n.

*Column 8 is the photon frequency of the photon corresponding to the entron of state n.

*Column 9 is the electron orbit frequency.

We are going to assume that the kinetic energy of an orbiting electron in state 7 is the ground state kinetic energy of the electron (i.e. 13.6057 eV) minus the entron energy in states 2 through 7. Subtracting the total of the numbers in column 2 (states 2 through 7) from 13.6057 eV gives us, for state 7, the kinetic energy of the electron that has absorbed the energy of all entrons in states 2 through 7. So the kinetic energy of the electron in State 7 is 0.2829 eV.

Subtracting the numbers in column 2 (2 through 8) from 13.6057 eV gives us for state 8 the kinetic energy of the electron that has absorbed the energy of all entrons in states 2 through 8. So the kinetic energy of the electron in state 8 is 0.2126 eV. As explained above all of these entrons are additive. These values are reported in column 4. The electron orbit diameters are calculated using the Professor Bohr formula as explained above in the section entitled "Energetic Electrons" and these values are shown in Column 5. The orbiting electron velocities are reported in Column 6. Entron diameters corresponding to the energy values in Column 3 are reported in Column 7. Photon frequencies corresponding to the photon energies on Column 3 are reported in Column 8. The photon frequencies are based the speed of light being about 3×10^8 m/s. Electron orbit frequencies corresponding to the electron velocities in Column 6 and the orbit diameters in Column 5 are shown in Column 9.

As explained above, if the electron in a ground state hydrogen-1 atom absorbs a 13.6057 eV entron of a 13.6057 eV photon the kinetic energy and velocity of the electron will be reduced to about zero and the orbit diameter of the orbiting electron will be increased (effectively) to infinity and the electron will become dissociated from its proton. The 13.6057 eV photon has a wave length of about 91.2×10^{-9} m (91.2 nm) and represents the highest energy, shortest wavelength spectral line in the Lyman Series of spectral lines. The frequency of a 13.6057 eV ultraviolet photon is about 3.29×10^{15} cycles per second (almost exactly one-half of the ground state electron orbit frequency). According to the Ross Model, the entron diameter of the 13.6057 eV photon is about 0.6368×10^{-10} m which is about 60 percent of the electron orbit diameter of the ground state electron in the hydrogen-1 atom.

Table XI-A
Lyman Lines

(1) n	(2) Total Photon Energy (eV)	(3) Delta Entron Energy Difference E_n - E_{n-1} (eV)	(4) Electron Kinetic Energy (eV)	(5) Electron Orbit Diameter $(10^{-10}$ m)	(6) Electron Velocity $(10^6$ m/s)	(7) Delta Entron Diameter $(10^{-10}$ m)	(8) Delta Photon Freq. $(10^{15}/s)$	(9) Electron Orbit Freq. $(10^{15}/s)$
1	0	0	13.6057	1.0584	2.1877			6.6595
2	10.2043	10.204	3.4014	4.2334	1.0938	0.8491	2.467	0.8224
3	12.0939	1.8896	1.5118	9.5248	0.7292	4.5851	0.4569	0.2437
4	12.7553	0.6614	0.8504	16.933	0.5469	13.099	0.1599	0.1028
5	13.0614	0.3061	0.5443	26.455	0.4376	28.304	0.0740	0.0526
6	13.2278	0.1664	0.3779	38.104	0.3646	52.067	0.0402	0.0305
7	13.3228	0.1010	0.2829	50.900	0.3155	85.781	0.0244	0.0197
8	13.3931	0.0703	0.2126	67.715	0.2735	123.24	0.0170	0.0128
9	13.4377	0.0446	0.1680	85.712	0.2431	194.26	0.0108	0.0090

10   13.4696   0.0319   0.1361   105.80   0.2188   271.60   0.0077   0.0066

	Photon Energy		En − En-1		Photon Wavelength
166	13.60520625				
167	13.60521215		59×10^{-7} eV		21 cm

Large n   13.6057   0.0000   0.0000)   ∞   0.0000   ∞   3.290   (0 X ∞)

Table XI-B
Electron's Constant Steps

n	de Broglie Wavelength $\lambda = h/p$ (10^{-10} m)	Angular Momentum rmv (10^{-35} kgm^2/s)	Inverse Velocity 1/v (10^{-6} s/m)	Angular Velocity rv (10^{-4} m^2/s)
1	3.3249	10.545	0.4571	1.157
2	6.6501	21.089	0.9142	2.315
3	9.9751	31.633	1.3714	3.473
4	13.3003	42.178	1.8285	4.630
5	16.6221	52.726	2.2852	5.788
6	19.9472	63.274	2.7427	6.946
7	23.0552	73.140	3.1696	8.029
8	26.5425	84.354	3.6563	9.260
9	29.9213	94.900	4.1135	10.418
10	33.2451	105.432	4.5704	11.575
Large n	∞	?	∞	?

If you plug an n of 2 in Equation 8, you will see that the first spectral line in the Layman Series is about 10.2043 eV which corresponds to a wavelength of 121.6 nm. This line identifies the photons as photons in the ultraviolet portion of the electro-magnetic spectrum. The photon has an energy of 10.2043 eV and a frequency of 0.40546×10^{15} cy/s. Its entron also has an energy of 10.2043 eV and the entron has a diameter of 0.8491×10^{-10} m and a frequency of about 1.765×10^{18} cy/s. The capture by the ground state electron in the hydrogen-1 atom of the 10.2043 eV entron will reduce the kinetic energy of the electron from 13.6 to 3.4014 eV, reduce its velocity to 1.0938×10^6 m/s and increase its orbit to 4.2334×10^{-10} m. This turns out to be a stable orbit (at least temporally) for the hydrogen-1 orbiting electron and forms the basis for the Balmer series. Similarly the capture by the ground state electron of a 12.0939 eV ultraviolet entron results in a stable orbit and forms the basis of the Paschen spectral series. The same is true for the Bracket, Pfund and Humphreys spectral series and all of the other hydrogen -1 spectral lines.

The Balmer Spectral Series

The electron kinetic energy in the second Lyman line is equal to the electron kinetic energy in the first Balmer Line. With the capture of the 10.2043 eV entron the orbiting electron now has a velocity of 1.0938×10^6 m/s and its kinetic energy is 3.4014 eV (with an orbit diameter of 4.2334×10^{-10} m). The 3.4014 eV electron can have its velocity and kinetic energy reduced to zero and its radius increased to infinity with the absorption of a 3.4014 eV entron of a 3.4014 eV photon corresponding to an ultraviolet wavelength of 364.6 nm (0.364 micron).

Table XI
Balmer Lines

n	Photon Energy (eV)	Energy Difference $E_n - E_{n-1}$ (eV)	Electron Kinetic Energy (eV)	Electron Orbit Radius (10^{-10} m)	Electron Velocity (10^6 m/s)	de Broglie Wavelength $\lambda = h/p$ (10^{-10} m)	Angular Momentum rmv (10^{-35} J)	Inverse Velocity (10^{-6} m/s)
2	0	0	3.4014	2.1167	1.093	3.3249	10.545	0.914
3	1.8897	1.8897	1.5118	4.7624	0.7293	6.6501	21.089	1.371
4	2.5511	0.6614	0.8504	8.4664	0.5470	9.9751	31.633	1.828
5	2.8572	0.3061	0.5443	13.227	0.4375	13.3003	42.178	2.286

This corresponds to the highest energy (shortest wavelength) Balmer spectral line. The lowest energy photon of the Balmer series is a 1.8896 eV visible light, bright red photon with a wavelength of 656.3 nm (0.656 micron). When the 3.4014 eV (kinetic energy) electron of the hydrogen 1 atom captures a 1.889 eV entron of a 656.3 nm photon the kinetic energy of the electron is reduced from 3.4014 eV to 1.5118 eV and its velocity is reduced from 1.0938×10^6 m/s to 0.7292×10^6 m/s and its orbit diameter is increased to 9.5248×10^{-10} m. (This 1.5118 eV entron has a frequency of 6.0×10^{13} cy/s and a diameter of 38.07×10^{-10} m.) This also turns out to be a stable orbit (at least temporally) for the hydrogen-1 orbiting electron. This 1.5118 electron then is the foundation for the Paschen Spectral series and can form the basis for the Paschen spectral series.

The Paschen Spectral Series

Now this electron with its velocity of 0.7292×10^6 m/s and its kinetic energy of 1.5118 eV (with an orbit diameter of 9.5248×10^{-10} m) can have its velocity and kinetic energy reduced to zero and its radius increased to infinity with the absorption of a 1.514 eV entron of a 1.514 eV photon having a wavelength of 8.18×10^{-7} m (0.818 micron). This corresponds to the highest energy (shortest wavelength) spectral line of the Paschen series. The lowest energy photon of the Paschen series is a 0.613 eV photon with a wavelength of 1875 nm (1.875 micron). When the 1.513 eV electron of the hydrogen 1 atom captures a 0.613 eV entron of a 1875 nm photon, the energy of the electron is reduced from 1.513 eV to 0.900 eV and its velocity is reduced from 0.7294×10^6 m/s to 0.5627×10^6 m/s with an orbit diameter of 64.0×10^{-10} m. This also turns out to be a stable orbit (at least temporally) for the hydrogen-1 orbiting electron and can form the basis for the Bracket spectral series.

The Bracket Spectral Series

Now this electron with its velocity of $0..5627 \times 10^6$ m/s and its kinetic energy of 0.900 eV (with an orbit diameter of 32.0×10^{-10} m) can have its velocity and kinetic energy reduced to zero and its radius increased to infinity with the absorption of a 0.900 eV entron of a 0.900 eV photon (with a wavelength of 1.38 micron). This corresponds to the highest energy (shortest wavelength) spectral line of the Bracket series. The lowest energy photon of the Bracket series is a 0.306 eV photon (with a wavelength of 4.05 micron). When the 0.900 eV electron of the hydrogen 1 atom captures a 0.306 eV entron of a 4.05 micron photon, the energy of the electron is reduced from 0.900 eV to 0.549 eV and its velocity is reduced from 0.5627×10^6 m/s to 0.328×10^6 m/s with an orbit diameter of 105×10^{-10} m. This also turns out to be a stable orbit (at least temporally) for the hydrogen-1 orbiting electron and can form the basis for the Pfund spectral series.

The Pfund Spectral Series

Now this electron with its velocity of 0.328×10^6 m/s and its kinetic energy of 0.306 eV (with an orbit diameter of 57.6×10^{-10} m) can have its velocity and kinetic energy reduced to zero and its radius increased to infinity with the absorption of a 0.306 eV entron of a 0.306 eV photon. This corresponds to the highest energy (shortest wavelength) spectral line of the Pfund series. The lowest energy photon of the Pfund series is a 0.1662 eV photon with a wavelength of 7460 nm. When the 0..306 eV electron of the hydrogen 1 atom captures a 0.1662 eV entron of a 7460 nm photon, the energy of the electron is reduced from 0.306 eV to 0.1398 eV and its velocity is reduced from 0.328×10^6 m/s to 0.205×10^6 m/s with an orbit diameter of 412×10^{-10} m. This also turns out to be a stable orbit (at least temporally) for the hydrogen-1 orbiting electron and can form the basis for the Humphreys spectral series.

The Humphreys Spectral Series

Now this electron with its velocity of 0.205×10^6 m/s and its kinetic energy of 0.1398 eV (with an orbit diameter of 75.2×10^{-10} m) can have its velocity and kinetic energy reduced to zero and its radius increased to infinity with the absorption of a 0.1398 eV entron of a 0.1398 eV photon. This corresponds to the highest energy (shortest wavelength) spectral line of the Pfund series. The lowest energy photon of the

Humphreys series is a 0.100 eV photon with a wavelength of 12400 nm. When the 0.1389 eV electron of the hydrogen 1 atom captures a 0.100 eV entron of a 12400 nm photon, the energy of the electron is reduced from 0.1398 eV to 0.0398 eV and its velocity is reduced from 0.205 X 10^6 m/s to 0.1138 X 10^6 m/s with an orbit diameter of 1447 X 10^{-10} m. This also turns out to be a stable orbit (at least temporally) for the hydrogen-1 orbiting electron and can form the basis for additional spectral series. This process continues for all spectral series, even those that have no special names.

Quantum Increases

As shown in **Table XI-B** for each of the many series of hydrogen spectral lines the energy of the photons that are absorbed and emitted increase in integer steps:

(1) the deBroglie wavelength, $\lambda = h/p = 3.3249$ X 10^{-10} m,
(2) the angular momentum, rmv = 10.545 X 10^{-15} J,
(3) the inverse velocity, 1/v = 0.5471 X 10^{-6} s/m, and
(4) the angular velocity, rv = 1.157 X 10^{-4} m^2/s

The question is, "Why is this?" My answer is I do not know. Maybe the reader has the answer.

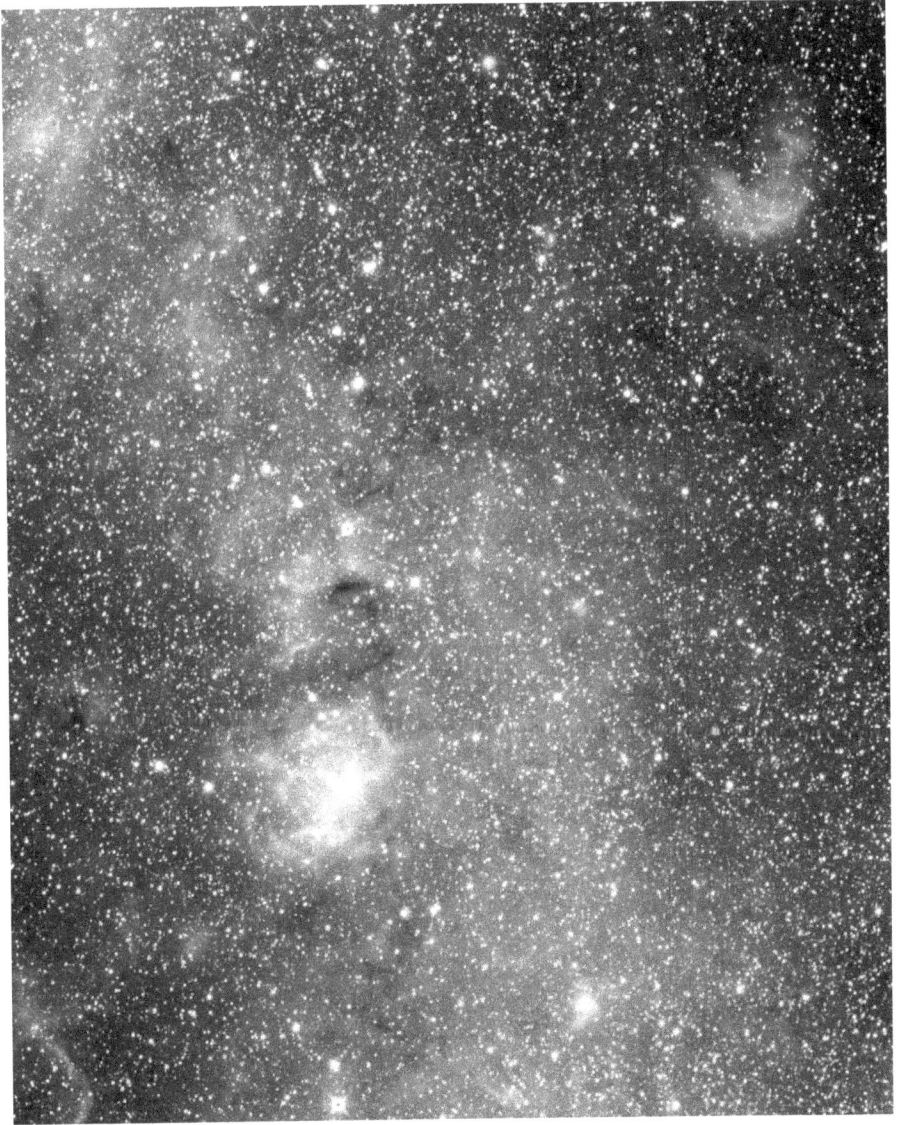

Trifid Nebula

Image credit: NASA/JPL-Caltech/UCLA

CHAPTER XXVII

EVIDENCE IN SUPPORT OF THE ROSS MODEL

Coulomb's Law

As explained in **Chapter III** Coulomb's Law requires all charged particles either to be point particles or be comprised of point particles, otherwise the charged particle would blow itself apart with repulsive forces. These point particles are tronnies.

Gravity

The speed of gravity is equal to the speed of light. Only photons travel at the speed of light. Therefore gravity must be carried by photons. The only photons that could carry gravity are photons that are able to penetrate massive objects with a few percent being captured by the massive objects for later release to provide the gravity of the massive objects. These photons are neutrino photons.

Destruction of Protons in Black Holes

Having neutrino entrons within protons that can be released by the destruction of the protons in Black Holes provides a simple explanation of galactic gravity.

Equivalence of the Numbers of Electrons and Positrons

We know that electrons can only be produced in pair production with the production of its anti-particle, the positron. And the only way to destroy an electron is to annihilate it with a positron. So there must be an equal number of electrons and positrons in our Universe. Only the Ross Model proposes this equality.

Electrons Are Self-Propelled

Only the Ross Model proposes that electrons are self-propelled at a very high speed. This is why electrons can orbit positively charged nuclei for billions of years without being drawn into the nuclei.

Coulomb Grids

The Ross Model provides an answer to the question of where charged particles get their charges. They get their charges from the speed of light Coulomb grids that fills our Universe. No other theory answers this question.

Anti-Gravity

The Ross Model provides a simple answer to the question of why far-away galaxies are all expanding away from each other and why close by galaxies are attracted to each other. In both cases the forces attracting and repulsing are provided by speed of light photons.

Dark Matter

Dark matter is the invisible neutrino photons that fill our Universe, each having a mass almost equal to the proton mass.

Wave Functions

The Standard Model defines the photon and the electron as wave functions. The Ross Model provides a simple description of these wave functions in which tronnies are points of focus of speed of light Coulomb force waves. So under the Ross Model, photons and electrons can be defined as wave functions. The difference between the Ross Model and the Standard Model is that the Ross Model shows how Coulomb waves can create particles with structure and mass. The Ross Model explains how tronnies can be formed into entrons and photons; how entrons can be combined to produce electrons and positrons; and how electrons, positrons and entrons can be combined to produce everything else in our Universe.

Relativity Theories

The Ross Model explains that massive objects, such as our earth, drags along its Coulomb grid. Light travels at the speed of light through these Coulomb grids. This is why the measured speed of light is a constant. The actual speed of light is not constant but depends on the direction and speed of the Coulomb grids through which the light is traveling. The Coulomb grid surrounding the massive objects is a composite grid consisting of a combination of the Coulomb grid produced by the massive object and the Coulomb grid associated with the space through which the massive object is traveling. If these curved Coulomb grids are defined as space, the massive objects will produce a curvature in that space. However, I do not believe most people would think of space as being defined by Coulomb grids. So my conclusion is that we should not promote an idea that massive objects curve space.

Empty Space

The Ross Model provides the best explanation of how our Universe could be created from empty space. The Ross Model proposes that our Universe is still 100 percent empty space with that empty space being completely filled with Coulomb grids created by point particles, tronnies, that occupy no space and are each the focus of Coulomb force waves. The Ross Model does not predict which came first, tronnies or the Coulomb forces they produce and are produced by.

Black Hole in the spiral galaxy NGC 3627

CHAPTER XXVIII

SOME SPECIFIC ROSS MODEL NUMBERS

Reference Numbers

Table VI provides some specific values describing features of the Ross Model.

Table VI
Ross Model Values

Entrons and Photons:

Ratio photon diameter d to entron diameter d': $d/d' = 911$

Photon wavelength: $\lambda = 1431d' = 1.5708d$

Photon energy E: $E = 1.389 \times 10^{-28} Jm/d' = 8.67 \times 10^{-10}$ eVm/d'

Electrons:

Mass of naked electron: $m_e = 9.109 \times 10^{-31}$ kg

Velocity of naked electrons: $v = 2.1877 \times 10^6$ m/s

Kinetic energy of naked electrons: $KE = 2.18 \times 10^{-18}$ J $= 13.6$ eV

Protons:

Mass of naked proton plus one naked electron:
$m = 0.99883825$ amu $= 1.658611 \times 10^{-27}$ kg

Velocity of naked proton v: $v = 4.02 \times 10^7$ m/s

Neutrino Photon and Neutrino Entron:

Wavelength: $\lambda = 1.336 \times 10^{-15}$ m

Diameter neutrino photon: $d = 0.85 \times 10^{-15}$ m

Diameter neutrino entron: $d' = 0.934 \times 10^{-18}$ m

Energy neutrino photon and neutrino entron: $E = 928$ MeV $= 1.487 \times 10^{-10}$ J

Mass neutrino photon and neutrino entron: $m = 1.65 \times 10^{-27}$ kg.

Here are 101 predictions of the Ross Model. We have suggested that you, the reader, take a few minutes to determine which if any of these predictions you believe will be determined are correct before you read the book and that you do the same after you have read the book. Readers are encouraged to try to prove these predictions incorrect or correct. If you can do either Mr. Ross would be happy to hear from you. He hopes to have his web site in operation soon at www.tronnies.com.

CHAPTER XXIX

PREDICTIONS OF THE ROSS MODEL

One Hundred and One Predictions

Tronnies

1. Tronnies are the basic building blocks of our Universe. Each one is a mass-less, volume-less point particle, each with a charge of + or - e, i.e. about +/- 1.602×10^{-19} coulomb.

2. Our Universe is comprised of nothing but tronnies and things made from tronnies.

3. The number of plus and minus tronnies in our Universe is equal.

4. Tronnies exist only in the form of entrons, electrons and positrons.

5. Each tronnies is a point focus of speed-of-light Coulomb forces.

6. Tronnies must travel in circles at a speed of $\pi c/2$ (about 1.57 times the speed of light) in order to exist.

Entrons

7. An entron is a plus tronnie traveling in a circle at $\pi/2$ times the speed of light combined with a minus tronnie also traveling at $\pi/2$ times the speed of light on the opposite side of the circle.

8. Entrons are the basic energy/mass quantum of our Universe.

9. All entrons are exactly alike except for their size, energy and mass; the largest entrons being about one thousand-trillion times larger than the smallest entron and the smallest entrons being about one thousand-trillion times more energetic and massive than the largest entrons.

10. Entrons are absorbed by electrons, positrons, protons, atoms and molecules increasing their mass in the process.

11. Entrons are absorbed in matter increasing its temperature.

Neutrino Entrons

12. The most energetic natural entron is the neutrino entron which has energy of about 9.28×10^8 eV.

13. A neutrino entron provides almost all of the mass of each proton.

14. Neutrino entrons represent most of the mass of our Universe.

15. A neutrino entron combines with an electron to produce a very high energy-mass electron which in turn captures two positrons to form a ground state

proton.

16. Protons are destroyed in Black Holes releasing its neutrino entron to produce the gravity of galaxies.

17. A neutrino entron combines with a gamma ray entron and a low-energy entron in a pair production process to produce an electron and a positron.

18. In electron-positron annihilation a neutrino entron is released as a neutrino photon.

Photons

19. Each photon is comprised of one entron.

20. The mass of the entron is also the mass of the photon.

21. The photons energy is proportional to its mass, $E = mc^2$.

22. In a photon the entron travels in a circle at a speed of twice the speed of light and forward at the speed of light.

23. The entron's circle within the photon has a diameter equal to about 0.6366 times the wavelength of the photon.

24. An entron's circle within a photon has a diameter of about 911 times the diameter of the entron.

Electrons and Positrons

25. There are an equal number of electrons and positrons in our Universe.

26. A ground state electron is comprised of three tronnies, one plus tronnie and two minus tronnies.

27. A ground state positron is comprised of three tronnies, one minus tronnie and two plus tronnies.

28. Ground state electrons and ground state positrons are self-propelled by internal Coulomb forces at a speed of about 2.19×10^6 m/s, a little less than one percent of the speed of light.

29. Most electrons orbiting in atoms are ground state electrons with zero electrical energy, self-propelled at a speed of about 2.19×10^6 m/s.

30. Ground state electrons capture low-energy entrons to slow down.

31. Ground state electrons capturing a high-energy entron can be driven at very high speeds by the captured entron.

32. However the high-energy entrons add to the mass of electrons which explains the mass increase of very high-speed electrons.

33. When an electron captures a neutrino entron its mass is increased almost to the mass of a ground state proton and Coulomb forces of the neutrino entron and other Coulomb forces drive the electron in a circle at a speed of $\pi/2$ times the speed of light.

Protons

34. Each ground state proton is comprised of two naked positrons and one very high-energy electron.

35. Ground state protons are self-propelled by internal Coulomb forces at a very high speed.

36. That very high speed is about 4.02×10^7 m/s (i.e. about 13 percent of the speed of light).

37. Ground state protons slow down by capturing gamma ray entrons to become hydrogen nuclei and once the hydrogen nuclei is moving slowly enough it can capture an electron to become a hydrogen atom.

38. A proton at speeds close to zero has captured a number of gamma ray entrons having a total energy of about 8.37 MeV (equal to about 1.341×10^{-12} J) which increases the mass of the proton by about 1.493×10^{-29} kg (about 0.9 percent).

39. Protons give up these gamma ray entrons when they combine in fusion reactions to form helium nuclei (alpha particles).

40. This energy is fusion energy, the energy of the hydrogen bomb and the energy of stars.

Neutrons

41. Neutrons are protons combined with a high-energy electron (i.e. a beta particle).

42. The high-energy electron is a naked electron with a captured gamma ray entron.

43. The mass of the naked electron plus the mass of the captured gamma ray entron represent the difference in mass between the proton and the neutron.

44. Neutrons are unstable with an average lifetime of about 15 minutes whether they are within the nucleus of an atom or outside the nucleus of an atom.

Atomic Nuclei

45. An alpha particle is comprised of four naked protons and two electrons plus some gamma ray entrons.

46. Alpha particles lose some of their gamma ray entrons each time they combine with other alpha particles and other particles to form heavier nuclei up to iron-56.

47. Nuclei heavier than iron-56 need to have captured additional gamma ray entrons to help hold the nuclei together.

48. There is no "strong force" holding nuclei together; Coulomb forces hold nuclei together.

49. There is only one basic force operating in our Universe, the Coulomb force produced by tronnie and things made from tronnies.

50. There is no such thing as quarks.

Gravity

51. Anti-protons are produced in Black Holes, each one by the combination of a high-energy positron (a neutrino entron and a positron) and two electrons.

52. Protons and anti-protons are destroyed in Black Holes when they combine.

53. Two neutrino entrons are released with each destruction.

54. Neutrino entrons escape the Black Holes as neutrino photons.

55. Neutrino entrons have a diameter of only about 0.934×10^{-18} m about 70,000 times smaller than an X-ray entron.

56. Most neutrino photons illuminating stars, planets and moons pass right through them.

57. Coulomb forces from the tronnies within the neutrino photons apply a tiny force on the charged particles of objects through which the neutrino photons pass.

58. The net force on the objects is in a direction toward the source of the neutrino photons, initially the Black Holes.

59. This net force is the force of gravity.

60. This explains why stars and their planets orbit Black Holes and are sometimes driven into Black Holes by this force of gravity.

61. Stars and planets driven into Black Holes provide protons that are destroyed to produce more neutrino photons.

62. Some neutrino photons are temporally stopped by particles within the objects through which the neutrino photons pass and are later released in random direction.

63. This is how stars, planets and moons get their gravity.

Anti-Gravity

64. Visible light and other low-energy radiation traverses inter-galactic space better than neutrino photons.

65. Anti-gravity is the result of photon pressure from the stars of galaxies on the stars, planets and moons of far-away galaxies; and neutrino photons from the shell of our Universe may also contribute to anti-gravity.

Galactic Attraction and Repulsion

66. Neutrino photon gravity effects trump low-energy radiation anti-gravity effects with respect to nearby galaxies and low-energy radiation anti-gravity

effects trump neutrino photon gravity effects with respect to far-away galaxies; therefore, close by galaxies are attracting each other and far-away galaxies are repelling each other.

The Speed of Light

67. Our Universe is completely filled with Coulomb force waves all traveling at the speed of light in all directions.

68. This creates a huge number of Coulomb grids, including a universal grid in which all other Coulomb grids within our Universe exists, with each grid being associated with a system of charges.

69. These systems include our entire Universe, each galaxy, each star and its planets, each planets and its atmosphere and much smaller objects such as glass balls, prisms and lenses.

70. Most Coulomb grids (such as the Coulomb grid of our solar system and the Coulomb grid of our earth and its atmosphere) are moving rapidly relative the universal Coulomb.

71. Light slows down when passing through a Coulomb grid which is heading in an opposite direction so as to travel at the speed of light relative to the Coulomb grid.

72. Light speeds up when passing through a Coulomb grid heading in a same direction so as to travel at the speed of light relative to the Coulomb grid.

73. Our earth drags its Coulomb grid through our Universe.

74. Instruments on earth designed to measure the speed of light all travel at the same speed as the earth's Coulomb grid, so we always measure the vacuum speed of light as the vacuum speed of light (i.e. about 3×10^8 m/s).

Electric Charge

75. Tronnies are the source of all electric charge.

76. Each tronnie possesses a charge e (about 1.602×10^{-19} coulombs).

77. Tronnies must always travel in circles at speeds of about 1.57 times the speed of light relative to a Coulomb grid while its Coulomb force waves are all traveling at the speed of light relative to the same grid.

78. Each tronnie therefore is continually at a focus of its own Coulomb forces coming at it diametrically across the diameter of its circle.

79. For example the two-tronnie's circle in a green light entron has a diameter of 3.77×10^{-10} meters (a little less than four tenths of a billionth of a meter) and each of the tronnies is circling at a frequency of about 4×10^{17} cycles per second.

80. As these Coulomb forces, coming from all points on the circle (360 degrees), pass through the tronnie (remember the tronnie is a point), they spread out in

all directions (360 degrees) from the tronnie's fast moving focus point.

81. Charge can be defined as Coulomb forces spreading out in all directions at the speed of light.

82. Therefore, tronnies get their charges at least in part from themselves.

83. However, each tronnie is required to be associated with at least one other opposite tronnies to assure that it travels in a circle.

84. Entrons are two dimensional and so are their photons.

85. This is why light can be easily polarized.

86. Electrons and positrons are three dimensional.

Magnetism

87. Magnetic fields are produced by naked electrons traveling through magnetic material at their natural speed of about 2.19×10^6 m/s.

88. It takes a few seconds for naked electron to make a loop from the earth's South Magnetic Pole through the earth to about the magnetic North Pole and then half way around the earth back to the magnetic South Pole.

89. In electric generators naked electrons (with the same charge as energetic electrons) from strong magnets moving relative to coils of copper wire force low-energy and high-energy electrons to flow back and forth through the copper wire to create alternating electric current in the copper wire.

Space

90. Assuming empty space traversed by Coulomb force waves is still empty space, our Universe is 100 percent empty space since everything in our Universe is made from tronnies that are points of focus of Coulomb waves.

91. Each of us is 100 percent empty space.

92. On the other hand, if you believe the tronnie is not a point particle, then you must be treating its Coulomb force waves as part of the tronnie, and if that is the case each tronnie completely fills our entire Universe with its Coulomb force waves leaving zero empty space in our Universe.

The Shell of Our Universe

93. Our Universe is contained in a cold plasma shell of mostly naked electrons and naked positrons.

94. The shell is expanding due to pressure from low-energy radiation coming from the stars of 100 to 400 billion galaxies in our Universe.

95. The shell reflects the low-energy radiation and since the shell has been expanding rapidly for the past 13 to 15 billion years the reflected radiation has been greatly Doppler shifted downward in energy creating the low temperature cosmic background radiation.

96. Neutrino photons from Black Holes react with a naked electron and two naked positrons to create naked protons in the shell, and the naked protons capture gamma ray entrons to become hydrogen atoms which accumulate to become new stars and galaxies at the outer edge of our Universe.

Recycling of Our Universe

97. Our Universe will continue its expansion for at least several billion more years due to photon pressure.

98. Sometime in the future the radiation pressure from stars will diminish, relative to the neutrino photon flux from Black Holes, to a point in time when the gravitation effects of the neutrino photons will cause all galaxies to attract each other and the shell surrounding our Universe to begin to contract.

99. The contraction will be slow at first then the attraction will accelerate and after a period of many billions of years from now, our entire Universe, except for a portion of its shell and some of its neutrino photon flux, will be forced into a last remaining extremely massive Black Hole.

100. All atomic matter including all protons, electrons and positrons will be converted into entrons of all energies up to neutrino photon energies.

101. This last remaining extremely massive Black Hole will then explode in a Big Bang marking the death of our Universe and the birth of our successor universe!

APPENDIX

THE DRAWINGS

BRIEF DESCRIPTION OF THE DRAWINGS

FIG. 1 is a graphical model of a tronnie.

FIG. 1A is drawing of an entron.

FIG. 2A is a graphical model of an entron showing integrated forces on its tronnies.

FIG. 2B is another graphical model of an entron demonstrating why it is so stable.

FIG. 2C and 2D show an entron traveling as a part of a photon.

FIG. 3 is a graphical model of a photon viewed from a frame traveling with the photon.

FIG. 4 is a graphical model of a photon viewed from a frame stationary with respect to the photon.

FIG. 4A is a graphical model of a neutrino photon demonstrating the force of gravity applied by the neutrino photon on a charged particle.

FIG. 5 is a graphical model of a naked electron.

FIG. 6 is a graphical model of an energetic electron.

FIG. 7 is a graphical model of a naked proton.

FIG. 8 is a graph showing electron and proton velocity as a function of the energy of its captured entron.

FIG. 8A is a first graphical model of a neutron.

FIG. 8B is a second graphical model of a neutron.

FIG. 9 is a graphical model of a deuteron.

FIG. 10 is a graphical model of a tritium nucleus.

FIG. 11 is a graphical model of an alpha particle.

ABOUT THE AUTHOR

Mr. Ross grew up in the small historic town of New Bern, North Carolina. He attended North Carolina State College (now North Carolina State University) where he graduated with a degree in Nuclear Engineering in 1960. He went to work as a Test Engineer at the world's first commercial nuclear power plant, Shippingport Atomic Power Station, near Pittsburgh, Pennsylvania and was put in charge of nuclear physics testing and taught reactor physics in a training program for future nuclear plant operators. While employed at the power company, Duquesne Light Company, he attended Duquesne University Law School at night and graduated in 1967. He then joined the Westinghouse Corporation Law Department in 1968 and was assigned to work with its Nuclear Energy System Division which was then selling nuclear plants like hot cakes. In 1975 he was recruited by General Atomic Company in San Diego, California, to help sell its nuclear power plants. However, the oil embargo of the early 70's, combined with Chernobyl and Three Mile Island put a halt to nuclear plant construction in the United States and as a result he became General Atomic's patent attorney. In 1986 he joined Western Research Corporation which was conducting Star Wars research for the United States government. In 1996 he was hired by Cymer Corporation which makes the excimer lasers used as the light source in the fabrication of integrated circuits and obtained for them more than 100 patents mostly covering various versions of their lasers systems. In 2003 he returned to the successor of Western Research, Trex Enterprises Corporation as Vice President Intellectual Property. Trex is still involved mostly in government supported research and development. He and his son, also a patent attorney, operate a small patent practice called Ross Patent Law Office from their home offices in Del Mar and San Marcos, California. He has obtained hundreds of patents for his employers and clients in a variety of fields including, lasers, electronics, communication, thermoelectric devices, automation, imaging and integrated circuits. In his spare time during the past eight years, he has realized the existence of tronnies and entrons and has developed this "theory of everything" that he calls the "Ross Model of our Universe". This theory is a work in progress. He calls this version "Version Eight". Seven earlier versions are documented in patent applications that he has filed over the past seven years. Most of them can be downloaded from the US Patent Office web site. Just go to www.uspto.gov and search for "tronnies".

ABSTRACT

Our Universe in a Nutshell

Everything in our Universe is made from tronnies or from things made from tronnies. Tronnies are point particles with no mass and no volume and a charge of plus e or minus e. Electrons are comprised of one plus tronnie and two minus tronnies. Positrons are comprised of one minus tronnie and two plus tronnies. Ground state electrons and positrons are self-propelled at a speed of about 2.19×10^6 m/s. Magnetism is the flow of these ground state (zero energy) electrons through and around magnetic material. Entrons are comprised of one plus tronnie and one minus tronnie each traveling on opposites sides of a circle at a speed of $(\pi/2)c$. The smallest entron is neutrino entron with a diameter of 0.934×10^{-18} m and an energy/mass of 9.28×10^8 eV and 1.65×10^{-27} kg. The largest entrons have an energy/mass in the range of about 1×10^{-8} eV and 1.78×10^{-44} kg. Ground state electrons and positrons can capture entrons to become energetic carrying electric energy. A photon is comprised of one entron that determines the photon's energy/mass. Entrons travel within the photons in circles at a speed of $2c$ and forward at a speed of c. The diameter of the photon circle is 911 times the entron diameter. Each anti-proton is comprised of a massive positron (with a captured neutrino entron) and two electrons. Each proton is comprised of a massive electron and two positrons. The hydrogen nucleus is comprised of a proton with several captured gamma ray entrons. Alpha particles are produced through the fusion of four hydrogen nuclei with the release of some of the gamma ray entrons of the hydrogen nuclei. The nuclei of atoms larger than helium are primarily combinations of alpha particles (three or more of which can be attractive to each other at very short distances). Nuclei also include gamma ray entrons and may also include one or more extra protons and one or/or more extra electrons.

A Black Hole is located at the center of each galaxy and, at temperatures much higher than the temperature on stars, continuously consumes a portion of its galaxy breaking down atoms to release their protons. Anti-protons are produced in the Black Holes. The anti-protons and protons annihilate each other releasing their positrons, electrons and their neutrino entrons. Some of the neutrino entrons escape from the Black Hole as neutrino photons to produce the gravity of the galaxy holding star systems in orbit and the gravity that allows the Black Holes to consume enough of its galaxy to supply the galaxy with its needed gravity. Most neutrino entrons pass through stars, planets and moons of the galaxies applying a backward force directed toward the Black Hole, but some are temporarily stopped and later released randomly as neutrino photons to produce the gravity of the stars, planets and moons. Low-energy photons pass through inter-galactic space better than neutrino photons. Anti-gravity is the result of pressure of low-energy photon illumination from far-away galaxies. Our Universe is currently expanding due to this photon pressure and will continue to expand for billions of years, but a Monster Black Hole is building up near the center of our Universe. Eventually, as the Monster Black Hole grows, all of the galaxies of our Universe will begin to feel the neutrino photon gravitational pull of the Monster Black Hole. Gravitational contraction toward the Monster Black Hole will continue for billions of years so that when the far-away galaxies arrive at the vicinity of the Monster Black Hole they will be traveling at speeds many thousand times the speed of light. Most of these galaxies will crash into the Monster Black Hole and be consumed by it. But the Big Bang that will mark the death of our Universe and the birth of our successor Universe will occur while many of these galaxies are still on their way. These galaxies will pass through the hot radiant remains of the Monster Black Hole and expand out in every direction at many thousand times the speed of light to provide the inflation period of our successor universe.

www.ingramcontent.com/pod-product-compliance
Lightning Source LLC
Chambersburg PA
CBHW030332220326
41518CB00047B/1977